Municipal water and wastewater reforms in the Russian Federation

Sergey Gennadyevich Ivanov

Sergey Borisovich Sivayev

Ella Stanislavovna Shalukhina

IWA Publishing

water21
magazine of the International Water Association

Published by IWA Publishing Ltd, Alliance House, 12 Caxton St, London, SW1H 0QS, UK
Tel: +44 (0)20 7654 5500. Fax: +44 (0)20 7654 5555. Email: publications@iwap.co.uk
Web: www.iwapublishing.com

This report is published as part of the Water21 Market Briefing Series. Water21 is the magazine of the International Water Association.

First published 2007

British Library Cataloguing in Publication Data
A CIP catalogue record for this book is available from the British Library

Library of Congress Cataloging- in-Publication Data
A catalog record for this book is available from the Library of Congress

ISBN: 184339135X
ISBN 13: 9781843391357

Contents

List of abbreviations:

CFD	Central Federal District
CHS	Central Heating Station
FAA	Federal Anti-monopoly Agency
FEFD	Far-Eastern Federal District
FTA	Federal Tariff Agency
MHE	Municipal Housing Economy
MUE	Municipal Unitary Enterprise
NWFD	North-West Federal District
RF	Russian Federation
SBFD	Siberian Federal District
SFD	Southern Federal District
UFD	Uralian Federal District
VFD	Volga Federal District
WSSF	Water Supply and Sewerage Facilities

List of Figures:

List of Tables:

1. Introduction

1.1 Preamble

In this early part of the 21st century, Russia (the Russian Federation / RF) is going through a complicated period of transforming its socialist economy into a capitalist one, a process that began at the end of the 1980s. In the new type of economy, the principle is that ownership of the means of production passes from the state to the private sector.

Reforming nearly all of the foundations of the country's legislation anew over a very short period was not a simple task - securing rights of private ownership, creating investment potential and, most importantly, beginning to change the psychology of the ordinary people, who had become used to the state taking responsibility for everything, not themselves, during the socalist period.

This report presents a brief analysis of the condition of the municipal water supply and sanitary systems in Russia in the early part of the 21st century and tries to provide answers to the eternal questions: what is the cause of the problems and what needs to be done in order to solve them?

The report will be of interest to those trying to understand the transformation processes taking place in Russia in the field of municipal water supply and wastewater disposal, and also to potential investors considering possibilities of long-term investments into management efficiency optimisation, reconstruction and development of engineering infrastructure in Russia taking advantage of the favourable investment climate created recently in the country.

1.2 About the authors

Ivanov Sergey Gennadyevich, candidate of engineering sciences and General Director of Company 'Vodokanal-Invest-Consulting', has worked for over 15 years as one of the leaders of the municipal water supply Vodokanal for Ufa city (population over one million), and has also worked for three years as the deputy project director of the World Bank 'Municipal Water Supply and Sanitary Works' project, on the 'Technical Assistance for the Implementation of Operation and Management Efficiency Optimisation Program' component.

Sivayev Sergey Borisovich, candidate of economic sciences and Director of Fund, 'Institute of Municipal Economy', has been involved in the development of the Russian Federation Housing Code, the federal law 'About the tariff management basis for services of municipal enterprises' and many other regulatory documents. He has also taken part, as a consultant and the head of consulting teams, in many Russian and international projects relating to diagnostics and the development of municipal water supply systems and sanitary works for various Russian cities.

Shalukhina Ella Stanislavovna, executive director of Company 'Vodokanal-Invest-Consulting', has an MBA degree and has worked as a organizational development consultant for the World Bank and the EBRD Municipal Water Supply and Sanitation sector projects, on the Management Efficiency Optimisation programme component. At the present time Ella Shalukhina takes part in a project for investment of capital in the water supply sector of Russia.

1.3 General information about the Russian Federation

1.3.1 Geographic location, territory and population

Russia occupies the majority of Eastern Europe and the northern part of Asia. Its territory is bordered by the Atlantic, Arctic and Pacific oceans. About 75% of its territory consists of flatlands and lowlands. Mountainous areas prevail in the south and east, and 41% of the territory is covered by forest.

Russia is the biggest state in the world, covering 17 million square kilometres (in comparison, Canada covers 10 million square kilometres; the US 9.4 million square kilometres; Australia 7.7 million square kilometres; Ukraine 0.6 million square kilometres; France 0.55 million square kilometres; Germany 0.36 million square kilometres; and Great Britain 0.24 million square kilometres). The length of Russia, from the west to the east, is about 9000km. The overall length of Russia's borders is 58,600km, including over 40,000km of sea borders.

Russia has borders with Norway, Finland, Estonia, Latvia and Lithuania to the north-west, Byelorussia and Ukraine in the west and south-west, Georgia, Azerbaijan, Kazakhstan, China and Mongolia to the south, the Korean People's Democratic Republic to the south-east and has sea borders with Japan and the US to the east.

The population of Russia as of the beginning of January 2006 was 143 million (2.3% of the world's population, putting Russia in sixth place in the world after China (1.3 billion population), India (1 billion), USA (280 million), Indonesia (230 million) and Brazil (175 million).

There are over 1,100 cities in Russia, 13 with populations of over one million, 20 with populations between 500,000 and one million and 120 with populations between 100,000 and 500,000. Russia is a multinational state (it has over 100 national and ethnic groups). The prevailing nationality is Russian (over 80%) and 73% of the population live in urban areas. The most widespread religion is Christianity.

1.3.2 Fresh water resources

Expert estimates suggest that Russia holds 20% of the world's total freshwater resources. Within its territory there are over 120,000 rivers with an overall length of 2.3 billion kilometres, and there are around three billion lakes. The majority of the country's rivers and lakes are covered with ice during winter.

Russia's biggest rivers are in Siberia, including the Lena (4,500km), the Amur (4,400km), the Yenisei, the Irtysh and – in the European part of Russia – rivers such as the Volga (3,500km) and the Don.

Russia also possesses significant underground water resources (both fresh and mineral water). A lot of its water is contained within glaciers, the biggest of which are found on Arctic Ocean islands. The mountain glaciers of the Caucasus and other Russian mountains are of great economic importance as they feed the country's rivers.

1.3.3 Constitutional formation and territorial administrative system

The Russian Federation is a democratic federal legal state with a republican government. The Federation consists of 88 equal constituent parts: 21 republics, seven territories, 48 regions, two federal cities (Moscow and St Petersburg), one autonomous region, and nine autonomous national areas. A complete list of the constituent parts of the Russian Federation is provided in Appendix 1. From 1 January, 2007 the Federation will have 86 constituent parts (as a result of three areas integrating: the Krasnoyarsk Territory, the Taimyr Autonomous Area and the Evenki Autonomous Area will merge as the Krasnoyarsk Territory).

State authority in Russia is exercised by the President, the Federal Assembly (the Federation Council and State Duma), the government and courts of the Russian Federation. The President of Russia is elected by direct secret vote by the whole of the country's population. The President serves a four-year term of office.

The constituent parts of the Russian Federation form their own governmental bodies. Local governments in Russia are rendered competent and empowered in accordance with the constitution of Russia and federal laws, though local government bodies are not included in the system of government authorities.

Republics included within the Russian Federation have their own constitutions, with the remaining elements of the Federation having charters instead. Each constituent part has its own legislation. The constitutions of Russia and its federal laws take precedence across the whole territory of the Russian Federation.

The constituent parts of the Russian Federation are formed into seven federal districts: Central, North-West, Southern, Volga, Uralian, Siberian and Far Eastern.

1.3.4 Demographic structure

The size of Russia's population increased until 1992, when it began to decrease (Figure 1.3.4). At

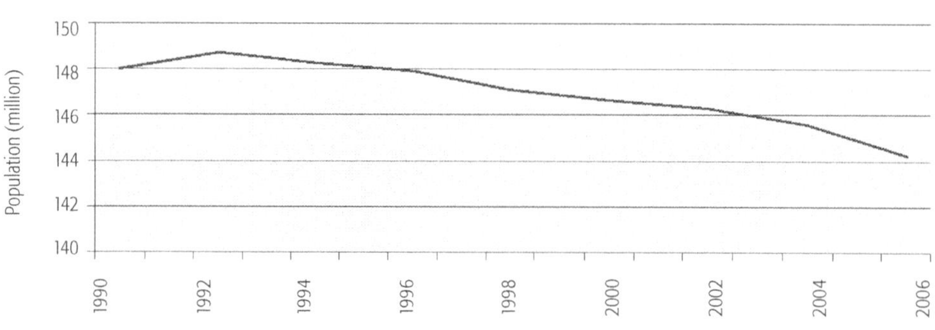

Figure 1.3.4. History of Russia's population fluctuations from 1990 to 2006.

present the population is decreasing by 600,000 to 800,000 people per year.

Using different forecasts (ranging from the pessimistic to the optimistic) by 2016 the population of Russia could be anything from 128 million to 142 million. The percentage of the population that does not work compared to the current working population is also predicted to increase from 54 to 68 per 100 people, that is, by 26%.

1.3.5 Timeline of the main social and economic indices in Russia from 2000 to 2004

During the period from 2000 to 2006 economic growth and high prices for energy products in the global market created significant wage growth, with consequent increases in the average income per capita and consumer expenditure (see Figure 1.3.5).

The number of people with an income below the minimum living wage reduced from 41.9 million in 2000 to 25.5 million in 2004, that is, by 64%.

Despite the reduction in the number of those considered to be living in poverty, because of inflation the wage gap with the minimum living wage is increasing. Between 2000 and 2004 this gap grew from 198 up to 227 billion rubles (from $7.7 billion to $8.8 billion).

The growth rate of public utility services is also outstripping that of inflation. During the five years from 2000 till 2004 public utility services in Russia grew on average by 3.5 times - the annual average growth of public utility service costs was 37% per year despite an inflation rate for that period of only 72% (15% per year on average).

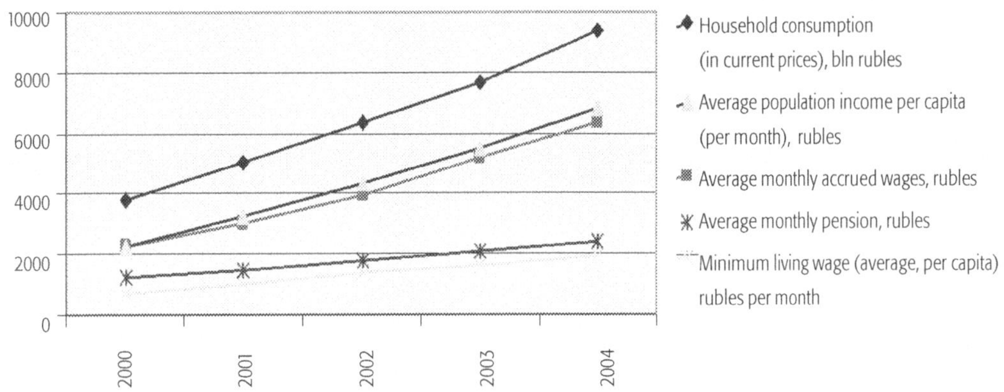

Figure 1.3.5. History of income and expenditure of the Russian people from 2000 to 2004.

2. Institutional issues

2.1 Background

Reforms planned by the Government of the Russian Federation envisage the implementation of measures aimed at promoting the attractiveness of the water supply and wastewater systems to private investors and fostering their participation to stimulate the sector's reconstruction and development.

The three-stage sub-programme entitled 'Reform and reconstruction of the municipal housing economy of the Russian Federation', approved by the Government in Decree 797 of 17 November 2001, sets out the problems facing the municipal housing economy (MHE) and provides perspectives on policy directions. The participation of the private sector in implementing the sub-programme is considered to be an important condition for financing the required investments (by 2010 attracting borrowed funding should become the main type of MHE financing) and attracting new management skills into the industry.

The public utility services element of the sub-programme is oriented specifically towards upgrading the water supply and wastewater disposal services undertaken by the municipal enterprises known as Vodokanals on behalf of consumers in the cities, settlements and villages of the Russian Federation. The main elements of the sub-programme that relate to water supply and wastewater disposal include:

- reconstructing the infrastructure
- reducing misallocations (for example, the misallocation of personnel, energy and water) by Vodokanals
- increasing the economic independence of the enterprises and their responsibility for the quality of customer service
- ensuring an optimal balance between the economic independence of the Vodokanals and state regulation of their activities, taking into consideration their monopolistic nature, the requirements for their development and the need for more powerful protection for consumer interests
- stimulating optimal economic development of the fixed assets and investments belonging to the Vodokanals by ending unnecessary investments and creation of fixed assets with excess capacity, perfecting water supply management and producing a reliable service

2.2 Objectives of Vodokanal reform

2.2.1 Objectives at federal level

Vodokanal reform at a federal level aims to initiate changes in the pattern of Vodokanal ownership that will facilitate acceptance of effective corporate management methods for the Russian municipal water supply and wastewater systems. It will set up a framework for creating the conditions to attract private sector resources and overcoming the many common problems that the entire industry faces. The main tasks of the federal government include:

- determining policy directions
- creating the coordinated legislative foundation needed to support the reform programme
- determining control mechanisms for Vodokanal activities (to be implemented by regional and local administrations and other engineering and economic regulatory authorities)
- appropriating funds from the federal budget to stimulate private sector support for the water supply and wastewater sectors in the poorest regions
- providing methodological support for the introduction of good corporate management principles at Russian water supply and sewerage facilities, and legislative support for good corporate management.

2.2.2 Objectives at Vodokanal level

Reform objectives at this level are:

- reforming the Vodokanals into legally-independent enterprises (joint stock companies), which act under the control of the municipal administration, or substituting Vodokanals for private sector operators, whose services are bought in a competitive bidding process, with an ultimate contract between the contractors and municipal administration setting out how to achieve a set of service level indices

• establishing formal relationships between the Vodokanal and the local administration; within the framework of such relationships the municipal administration can set objectives for the level of service and the Vodokanal will be granted a guaranteed income in perpetuity and autonomous management
• supporting the achievement of financial self-sufficiency and increasing the Vodokanals' borrowing capacity
• ensuring parallel access to investment funds from government programmes and the private sector to make up for the lost years when the industry suffered a lack of investment, technical maintenance and reconstruction.

In order to achieve the objectives of these reforms the municipal authorities should determine:
• a framework for provision of the required level of independence, for example by delegation, transferring Vodokanal functions to private concessions and liquidating the Vodokanal, through formation of a Municipal State Enterprise, or by corporatising the Vodokanal
• procedures and terms for Vodokanal reform
• an acceptable framework, strategy and policy for private sector participation
• measurement, design and monitoring of Vodokanal service level indices
• how to integrate post-reform schemes to monitor the economic and engineering indices and regulate the new Vodokanals

Related to this, the present guidance provides suggestions about patterns of ownership, structures of regulation and management for consideration by local administrations and their Vodokanals. If these structures are adhered to, they can form the basis and environment for Vodokanals to undertake the effective (and visible) performance of their main functions, that is, providing a safe and reliable water supply and wastewater disposal services for consumers at an optimal price.

2.2.3 Issues to be solved in the reform process

Before reviewing reform issues and discussing the possible involvement of the private sector it is necessary to stress that changing the Vodokanals' ownership pattern is not an end in itself, it is a mechanism that enables solutions to be found for a number of complicated current issues.

a. Administrative issues

Municipality administrations are responsible (at least in part) for the economic and technical regulation of Vodokanal activities, but some administrations fulfil these functions in a non-systematic and quite illogical way in terms of the public interest.

Relationships between municipal administrations and their Vodokanals in most cases are not clearly defined, and there is no impetus to set up a more transparent system. Some types of service agreement between Municipality administrations and Vodokanals have existed for a number of years, but most municipal administrations do not use them in practice.

The General Director of the Vodokanal is appointed by the city administration, that is, by the owner of the organisation. The appointment process is not always transparent. As a result, the General Director listens primarily to the city administration and other stakeholders inevitably receive less attention. Such conditions provoke conflicts of interest (between legal entities, political bodies and individuals) and set aside the public interest.

The incomes of the Vodokanals are dependent on an annual procedure of Municipality budget approval, which is why the attitude of the Vodokanal towards financial planning is reminiscent of a non-commercial, state-financed organisation (despite the fact that the accounting policy of the Vodokanals is based on the same principles as the policy for profit-making organisations).

Vodokanals do not work directly with their residential users, because in many cases payment collection is undertaken by the Municipal Services division, which acts as an intermediary.

For Vodokanals, having the status of municipal unitary enterprises means that the concepts and principles of corporate management in terms of protecting shareholders' rights are irrelevant in reality, but general principles of good management are nevertheless required (transparency of information). After delegating the Vodokanal's functions to a private operator through a concession contract or after corporatisation, the Vodokanals will have to comply with the requirements of the Russian code of good corporate management. These requirements will be included into business operation procedures from the very beginning.

b. Managerial, organisational and personnel issues

Some municipal administrations pull strings to influence the Vodokanal management and to pursue their own political objectives, despite the consequences their actions can have for water quality. Some Vodokanal managers do not have the motivation to raise the level of their own training, because there is no ability within the Vodokanals to manage the Municipal Unitary Enterprise (MUE) Vodokanal.

The level of qualifications achieved by many leaders and specialists in the Municipality administrations is not sufficient to enable them to accept literate, strategic solutions for the development of the municipal water and wastewater systems. At present the demand for the necessary skills is not great, but in the event of the Vodokanal reforms, the demand for specialists in management and strategic planning will increase.

There are not many educational institutions in Russia where refresher courses could be organised for the managers of water supply and sewerage facilities (WSSF). Large Vodokanals (for example, those in Moscow and St Petersburg) have created their own training centres. Exchange of advanced methods and technologies within Russia's WSSFs is undertaken at annual and bi-annual exhibitions and conferences, which attract considerable interest and can serve as a source of useful information for participants, as can specialised trade journals and publications. Unfortunately, Vodokanals do not have sufficient funds to finance the participation of significant numbers of their employees in training programmes.

c. Financial and investment issues

Municipal Services (the organisations responsible for billing and payment collection for public utility services) do not provide full payment collection for the services rendered by Vodokanals. Legal enforcement procedures for overdue amounts are practically impossible to uphold and can lead to Vodokanals going into bankruptcy.

Vodokanals do not receive full repayment for the services they provide. After the services they provide are valued this situation will be corrected, but the amount of uncollectable debt may increase.

The majority of local budgets receive government grants, and the Municipality administrations solve funding problems in their budget by redirecting the sums due from the public to Vodokanals for water supply and sewerage services not to the account of the Vodokanal, but to the account of an intermediary – Customer Service - after which the funds received for these services can be partially used for municipal administration requirements.

The amount paid for their services firstly often does not allow Vodokanals to attract credit funds to finance investments and repay loans, and secondly is insufficient to cover the operational costs of the Vodokanals' core activities. Under Federal Law No. 210- FZ of December 30, 2004, the tariff setting procedure was slightly liberalised, but the law has not yet been adopted and it is possible that the tariff setting procedure will still be used to limit tariff growth.

The annual tariff setting procedure is undertaken on the basis of an 'own costs plus a set limit of profitability' (cost plus) method which does not provide any motivation to reduce costs. If a Vodokanal is able to save some funds after tariff correction it can lead to a decrease in the tariff and income level and subsequent approval of new tariffs. A cycle of one year is too short to accumulate any funds for investment financing.

There is a tariff cross-subsidy for one group of consumers (the population) from other groups (other consumers). This cross-subsidy is gradually being stopped, but with socio-economic reasons guiding the situation, industrial consumers in many places are still faced with higher tariffs to protect the general population.

Bills for water supply and sewerage disposal services are rarely based on measured consumption, a situation that does not encourage consumers to reduce their consumption and save water. Deductions for depreciation have remained at the disposal of the Vodokanals since 1988, but they either could not (due to consumers' failure to pay) or did not want to use these funds for investment. Accounting controls for Vodokanals do not foresee a depreciation deduction analysis and a Vodokanal is entitled to use these funds at its own discretion. This is why, if a Vodokanal is a municipal unitary enterprise it can, if necessary, use depreciation deductions to cover its operational costs without coordinating with the owner of the water supply and sewerage system.

It is assumed that the Vodokanal's expenses, which are not included within the net cost of services when calculating tariffs, will be covered from income. Income, envisaged by tariffs for the Vodokanal's services, is not included in the net cost of services. At a federal level the maximum income is limited by the Russian Government to 25% of the net cost of its services, though the actual income value in a Vodokanal's tariff is set at a significantly lower level by the municipal administration and tariff control authorities.

The income envisaged in service tariffs is insufficient to enable the Vodokanals to undertake investment financing out of their own funds. Income (if any is received) is used to cover the difference between amounts owed under outstanding invoices and payments received. Overwhelmingly, Vodokanal profits are set at operational break-even for policy reasons. Investment-related expenses are usually not included in tariffs. Tariff control authorities justify this by explaining that the Vodokanal has, for these purposes, the two other sources of funding mentioned above. Federal Law 210 FZ envisages reimbursement of investment expenditure, but it does not define investment expenditures, and tariff growth will controlled by the tariff approval procedure.

Despite the fact that funds for refurbishment are included in the tariff calculation there is no clear reconciliation procedure between the municipal administration and the Vodokanal in terms of the size of these funds and the procedure for controlling expenses against planned values. This means there are no objective criteria for justifying the expenditure needed by the Vodokanal to undertake a complete refurbishment. That is why the Vodokanals' suggested expenditure for refurbishment is usually considerably reduced during the tariff review.

It is difficult to determine a Vodokanal's creditworthiness, as most Vodokanals do not use budgeting systems, financial modelling or forecast systems. The Vodokanals' accounting policy does not envisage openness in terms of information provision or the possibility of controlling finances (both the Vodokanal and the municipal administration may be interested in preserving this situation).

Vodokanals have little chance of attracting borrowed funds and as a result their attraction as an investment is very low. Banks do not give long-term credits to Vodokanals or if they do grant them, it is at a very high percentage interest that seriously limits the possibility of financing investment in infrastructure development. Federal Law 210- FZ sets up tariff rate control mechanisms for MHE enterprises to attract private sector investment, but the law does not give a precise definition of these mechanisms. A lot will depend on the tariff regulatory authority understanding what is required to attract investors.

d. Technical issues

The considerable wear and tear suffered by the Federation's infrastructure and equipment has led to unreasonably high water losses, low efficiency, water supply failures and a deterioration in water quality. There is some scope for urgently-needed works, which is why consumers have a low opinion of the quality of service provided by the Vodokanals.

This situation is predicated by the scarcity of funding available for financing investments over the past 20 years and the insufficient financing available for refurbishments and technical maintenance, funding often being sufficient only for emergency works. The simplicity of the water supply systems' operation and their managements' lack of technical skills are not the reasons for the bad state of the infrastructure.

For instance, the sector suffers from undue power usage, caused by earlier planning and design excesses that resulted in the construction of low-efficiency wastewater treatment facilities and pumping stations with too much capacity.

Insufficient metering of water consumption in the residential sector has also led to irrational over-consumption. This is exacerbated by the unwillingness of the Vodokanals' owners to support and implement federal resolutions or recommendations and the suggestions of the Vodokanals themselves relating to metering of water consumption.

Unsuccessful investment solutions are also common, such as the tendency to invest in construction of new production facilities, when water management measures could eliminate the need for such investments. The main reason why investment solutions fail, however, is the absence of action on the part of the Vodokanal's owners to create a real motivation to develop adequate feasibility studies for investment plans.

It is vital to understand that transforming the Vodokanals' corporate structure will not solve all issues listed above. For example, such transformation will not be able to solve:

- administrative issues, though it will create the background for their solution

• financial issues, though it will contribute towards stricter discipline in financial records
• investment issues, though it will open access to different mechanisms of debt financing (debt instruments and bonds) and will provide access to other sources of investment financing
• technical, management, organisational and personnel issues, but it will create an environment that will make it easier to solve these, if solving administrative and financial issues is prioritised.

The corporatisation of the Vodokanals should therefore be considered as a factor in creating the conditions for positive results from the reforms. The change of corporate form will create possibilities and conditions for implementing reform and attracting the private sector that are not provided by municipal unitary enterprises.

The corporatisation of the Vodokanals will also require fundamental changes in the relationship between the Vodokanals and the administrators of the Municipalities and this will lead, in turn, to a solution for the issue of the changes needed within the Vodokanals and municipal administrations.

2.3 Ownership and relationships between Vodokanals and their stakeholders

2.3.1 General information

The overwhelming majority (over 90%) of Vodokanals in Russia are municipal or state unitary enterprises. At present, their corporatisation is actively taking place.

Modified Vodokanals can be divided into four main groups:
• those with 100% Municipality participation in the authorised capital of the Vodokanal
• those where a controlling block of Vodokanal stock (50% + 1 share) is in the ownership of the Municipality;
• those where the share participation of the Municipality in the authorised capital of the Vodokanal is between 25 and 50%;
• those where the Municipality has no share in the authorised capital of the Vodokanal

Different management strategies are applied to each of the above cases.

2.3.2 Title to assets of municipal water supply and wastewater disposal systems

Federal Law No 178- FZ of 21 December, 2001 relating to state and municipal property does not exclude the sale of infrastructure to the private sector. Nevertheless, authorities at federal, regional and local levels have decided that title to infrastructure assets (utilities and facilities) should remain the property of the Municipalities.

Russian Federation legislation places organisational issues relating to water supply and sewerage within the jurisdiction of the Municipalities' competent administrations, which are fully responsible for these issues. Russia's civil legislation also foresees the responsibility of local authorities to provide the operation of vital assets in case of a Vodokanal's bankruptcy.

The following main principles are applied to the ownership of the water supply and wastewater systems:
• property used directly in operations, such as networks, facilities and process equipment (the first group), will remain municipal (state) property
• property for maintaining process equipment (the second group) (production and administrative buildings outside the boundary of the assets referred to as property of the water supply and sewerage systems, and all movable assets) can either remain municipal (state) property or be transferred to the private sector
• property belonging to the water supply and sewerage systems can be transferred to the Vodokanal to manage, under a lease or concession. If required, other forms of private sector engagement can be utilised for the operational management of the water supply and sewerage systems, provided that these forms comply with current federal laws.

2.3.3 Separation of ownership and operational responsibility with regard to water supply and wastewater networks

The following assets will remain municipal property:
• mains water distribution systems, including any wells within them, up to the flange after the valve in the flow at the point in the roadway where the customer service pipes connect to a central heating station (CHS) or an apartment building (transferred by the administration of the city under lease agreement to a Vodokanal)

• service pipes into the CHS (served by a heat supply company or by a Vodokanal under agreement with a heat supply company)
• main and sub-main sewerage systems, starting from the first well at the outlet from a residential block or CHS (transferred by the administration of the city under a lease agreement to a Vodokanal)
• hot and cold water distribution systems from the wall of the CHS building to the last valve including the well installed at the customer service pipe leading to a residential block (served by a heat supply company or by a Vodokanal under agreement with a heat supply company)
• wastewater outlets from a CHS (served by a heat supply company)

The following assets can be jointly owned by residential block owners:
• customer service pipes to residential blocks and sub-main distribution systems, including wells within them, up to the flange after the valve at the point in the network where the customer service pipe enters the street or sub-main water distribution network
• wastewater outlets, from the wall of the first well in a residential block
• water distribution (or wastewater) networks and standpipes inside a dwelling from the inlet (outlet) to their connection to the water distribution (wastewater) networks inside an apartment.

Water distribution and sewerage networks in joint ownership are serviced by the owners according to an agreement between them or by an arrangement in which the residential block becomes the management or where a contractor is hired by this organisation.

The following assets will be in private ownership:
• water distribution systems inside apartments from the connection to the standpipe and sewerage systems inside apartments from their connection to the soil stack
• individual (apartment) water meters, plumbing fittings and pump fittings

Water supply and sewerage systems in private ownership are serviced by the owners themselves or by a management company, or by a contractor hired by this company.

2.3.4 Definitions of a consumer, subconsumer, contractor and user

'Rules of use for municipal water supply and sewerage systems in the Russian Federation' (approved by the RF government on 12 February 1999 No 167) give the following definitions:
• consumers and sole traders are legal entities having ownership, operating control or operational management of assets, water supply and/or wastewater systems that are directly connected to public water and/or wastewater systems, having concluded water supply and/ or wastewater disposal contracts with a Vodokanal according to the established system.
• consumers can also be organisations that have housing associations and own engineering infrastructure assets, with operational control or management; organisations authorised to provide public utility services to the population of the state (department), municipal or public housing associations and other owners' associations that have the right to manage housing associations
• sub-consumer: an entity also known as a 'consumer' under the present rules, which receives drinking water from the water distribution systems and/or discharges wastewater into the wastewater network of a Vodokanal consumer.

The rules for provision of public utility services (approved by Resolution of the RF government on 26 September 1994 No 1099 regulate relationships between contractors and users of heat, power and water supply services (from this point, 'services') and provide the following definitions of contractor:
• enterprises or establishments having housing funds and owning engineering infrastructure assets, having operational control or management and duties, which include public service provision or enterprises and establishments authorised to undertake the functions of a contractor for consumers in a state (department), municipal or public housing associations
• an enterprise providing public utility services for a consumer in a private housing association
• condominiums, partnerships and other owners' associations, who have the right to residential block management, including the conclusion of service agreements and provision of public utility services for consumers living in the residential association as common property.

It also defines a user as: a citizen using or intending to use services on the basis of an employment agreement, lease agreement or accommodation servicing agreement finalised with a contractor

2.3.5 Regulation of technical, legal, financial and economic activities of a Vodokanal

a. Operational issues for centralised water supply and wastewater disposal systems

The Vodokanal is obliged:

• to operate a central water supply and wastewater disposal systems transferred to it in accordance with the 'Operational rules of municipal water supply and sewerage systems' (MDK 3-02.2001)

• to provide a set quality of drinking water to the city's distribution system and to consumers, complying with the requirements of the Sanitary rules and norms, SanPiN 2.1.4.559-96 'Drinking water. Hygienic requirements for water quality in central drinking water distribution systems. Quality control'.

• to ensure sufficient water pressure in the water distribution systems at the point where consumer service pipes connect, as set out in specifications issued to consumers

• to standardise and control the amount of drinking water taken by consumers from the central drinking water distribution system

• to standardise and control wastewater volumes discharged by consumers into the central sewerage system;

• to provide immediate notification through the mass media to drinking water consumers if pollutants are detected in concentrations hazardous to health

• to provide a central point of delivery for drinking water in tanks or other reservoirs to places of consumption and to arrange intake points to ensure drinking water is delivered to the population if pollutants are detected in the municipal distribution system in concentrations hazardous to health

The Vodokanal is entitled to:

• implement technical supervision of assets constructed by consumers to ensure their compliance with the specifications for connection of these assets to the municipal water supply and sewerage systems and the Vodokanal-approved design of asset connection to the municipal water supply and wastewater systems

• to implement technical supervision and control the correct operation by consumers of water supply and wastewater disposal systems entrusted to them to prevent entry through these systems of discharge-prohibited pollutants or concentrations of pollutants above the permitted level into central water supply or wastewater systems

• to undertake inspections of the operation of water supply and wastewater disposal systems entrusted to consumers to prevent unauthorised connections to the central water supply and wastewater disposal systems and unauthorised use of these systems

Any legal entity or person is entitled to claim compensation in relation to damage sustained in connection with obligations that a Vodokanal has not fulfilled.

b. Issues of consumer asset connections to central water supply and wastewater disposal systems

The Vodokanal is obliged to:

• at the request of any legal entity or person within 15 days (Clause 21 of the Rules of use of the municipal water supply and wastewater systems of the RF) after receipt of such a request, to issue specifications for the connection of assets controlled by such an entity or person to the central water supply and wastewater disposal systems or to prepare within the same period a justified refusal to issue specifications. When issuing specifications to a person for the connection of assets not related to their accommodation (either completely or partially), the Vodokanal is entitled to demand proof of this person's registration in the capacity of an entrepreneur (Clauses 1 and 3 of the Rules of use of municipal water supply and wastewater systems of the RF and Clause 1.1 of the Rules of public utility service rendering), and if such proof is not provided the Vodokanal is entitled to refuse to conclude the contract

• to review a project submitted for approval for connection of a consumer's (customer's) asset to the central water supply and wastewater disposal systems, to approve this project or direct to a consumer (customer) a justified refusal to approve the project (the length of time for consideration and approval of the project is not set in legislation, but is decided according to business practice and the time normally required for review and approval procedures)

• to issue a permit to a consumer (customer) for connection (insertion) of networks belonging to the consumer (customer) to the water supply or wastewater networks of the central water supply and wastewater disposal systems, to provide supervision of works undertaken by a consumer (customer) relating to network connections and measures for disinfection of the connected pipeline section until a positive result is received from the bacteriological analysis
• to supply water to a consumer and receive their wastewater through new connections according to conditions determined in their water supply and water disposal contract after the conclusion of such a contract or by the introduction of data about the new connection to a current contract

The Vodokanal is entitled:
• to refuse to issue to a consumer (customer) specifications for connection to the central water supply and wastewater disposal systems if they have no engineering capability to undertake such a connection or if this connection affects the network operation so that the Vodokanal will not be able to fulfil its water supply and wastewater disposal contract conditions agreed with other consumers
• to refuse to issue a consumer (customer) a permit for connection to the central water supply and wastewater disposal systems if a consumer (customer) violates or does not fulfil specifications for connection or undertakes connection works without approval

Any consumer (customer) is entitled to appeal to a court if the Vodokanal refusal to connect their assets.

c. Issues relating to the conclusion, modification and cancellation of water supply and water disposal contracts
A water supply and water disposal contract by virtue of Clause 11 of the 'Rules of use of municipal water supply and sewerage systems of the RF' and Article 445 of the 'Civil Code of the RF' refers to public contracts. In connection with this, the following concluding procedure will be applied to such contracts.

According to Clause 1 of this Article, a party wishing to conclude a contract (the consumer) will send to the other party, for whom the contract's conclusion is mandatory (the Vodokanal), a request issued according to Clause 12 of the' Rules of use of municipal water supply and sewerage systems of the RF'. This request will be considered as an offer submitted to the Vodokanal. The Vodokanal will review it within 30 days of receipt and will send to the consumer (customer) of the water supply and wastewater disposal project the conditions specified in the offer or will suggest its own conditions for the project, or will direct a justified refusal to conclude such contract. A draft contract directed by the Vodokanal to a consumer will be considered to be a new offer.

After receiving a draft contract from the Vodokanal, the consumer will review it and within 30 days of receipt the consumer will send either a signed contract or a signed contract with an attached discrepancy report.

If a discrepancy report is received, the Vodokanal will review it within 30 days of its receipt and notify the Consumer regarding its acceptance or rejection. The concerned party (the Vodokanal or the consumer) is entitled to present disputed matters to arbitrage if no agreement is reached.

Omitting the 30-day period set in Article 445 of the Civil code of the Russian Federation for transferring a discrepancy report to arbitration is not sufficient reason for a court to reject a petition to judge a pre-contractual dispute (Clause 1 of the Enclosure to the Information Letter of the RF Superior Commercial Court of 5 May 1997 No 14 Review of conflict resolution practice related to the conclusion, modification and cancellation of contracts).

The above period should not be considered as limiting the possibility of a concerned party presenting issues disputed under the contract to arbitration. That is why, in cases where the concerned party presents disputes to a court after the period expires and the other party does not object, the court can adjudicate a claim on its merits. If a concerned party does not present a discrepancy report about issues relating to essential contract conditions within 30 days, the contract will be considered unconcluded.

While taking a decision about whether discrepancies contained in a discrepancy report are essential conditions, it is necessary to be guided by Article 432 of the Civil code of the Russian Federation which states that essential conditions in the subject matter of a contract are those conditions named in law or in other regulatory documents as essential, and also all conditions that have to be agreed on by statement of any party. In accordance with Clause 13 of the 'Rules of use of municipal water supply and sewerage systems of the RF', a contract will contain subject matter, relating to water supply and/or wastewater disposal, in which the following essential conditions will be considered:

- the supply (receipt) mode of the drinking water, including the water supply mode during fire-fighting activities and receipt (discharge) of wastewater
- limitations to the supply (receipt) mode of drinking water and receipt (discharge) of wastewater
- the quality of the drinking water and regulatory requirements for wastewater composition
- conditions for termination or limitation of the drinking water supply (receipt) and wastewater receipt (discharge)
- accounting for supplied (received) drinking water and received (discharged) wastewater
- the settlement procedure, terms, tariffs and payment terms including for quantities in excess of the water consumption limit and quantities in excess of the wastewater and pollutant discharge limit
- the borders of the operational responsibilities of the parties for the water supply and sewerage networks
- the rights and obligations of the parties
- any penalty (charge, fine) and other types of deterrent envisaged in Russian Federation legislation for the non-fulfilment of contract conditions and for not properly fulfilling obligations
- other conditions, which should be agreed upon by a statement from any party

If the Vodokanal (the party for which contract conclusion is mandatory) presents discrepancies to a court and the consumer submits their previous suggestions for contract conditions, then arbitration will proceed on the basis that the dispute was presented by agreement of the parties and will adjudicate on its merits.

A claim petitioning conclusion of a public contract can be only issued by the party to a contract that must conclude it, that is, only by the consumer.

In arbitration proceedings practice the question arose of whether a commercial organisation, whose duties include undertaking work or services for each person that has requested such work and services, is entitled to demand the conclusion of a contract with a consumer if the latter declines to conclude it.

It follows from Clause 2 of Article 421 of the RF Civil code contract conclusion that petitioning is not allowed apart from in cases where the obligation to conclude a contract is envisaged by the present Code, law or in a voluntarily-undertaken obligation.

In accordance with Clause 1 of Article 426 of the RF Civil code, a public contract will be acknowledged as a contract concluded by a commercial organisation, which sets forth its obligations in regard to the sale of goods, undertaking of works or rendering of services, which by the nature of its activities such an organisation must undertake for each person that requests such goods, works or services.

Clause 3 of the same Article stipulates that if a commercial organisation unreasonably declines to conclude a public contract then a provision in Clause 4 of Article 445 of the RF Civil code will apply. Clause 4 of the specified Article provides that if a party, for whom conclusion of a contract is mandatory under the present Code or other laws, declines to conclude the contract, then the other party will be entitled to appeal to court with a claim to petition to conclude the contract.

By implication of Clauses 1 and 3 of Article 426, and also Clause 4 of Article 445 of the RF Civil Code only a contracting party to the obliged party can appeal to court with a claim petitioning for contract conclusion. A commercial organisation is not entitled to canvass a consumer to conclude such a contract.

In cases where a consumer uses water supply services provided by the Vodokanal (the obliged party), but nevertheless refuses to conclude the contract, it is necessary to bear in mind the following. Actual use by a user (consumer) of the services of the Vodokanal (the obliged party) in accordance with Clause 3 of Article 438 of the RF Civil code will be considered as acceptance of the version of the contract suggested by the Vodokanal. In this case, these relationships will be considered as contractual relationships according to the conditions offered by Vodokanal (Clause 2 of the Enclosure to the Information Letter of the RF Superior Commercial Court of 5 May 1997 No 14 Review of conflict resolution practice related to the conclusion, modification and cancellation of contracts).

d. Issues of rationing and accounting for water consumption from a central water supply system into the wastewater disposal system

The local authority, or the Vodokanal on local authority instruction, will set limits for water consumption and wastewater disposal (Clause 58 of the 'Rules of use of municipal water supply and sewerage systems of the RF') with consideration of the following:

- the actual capacity of the water supply and wastewater systems
- priorities in satisfying the demand for drinking water and wastewater disposal
- observation of limits for drinking water abstraction and wastewater and pollutant discharge into water bodies, as set by water and sewerage facilities, water management and environmental authorities
- the success of measures to rationalise drinking water use, and reduce wastewater and pollutant discharges by consumers;
- a consumer's water supply and wastewater disposal balance

In accordance with Clause 32 of the 'Rules of use for municipal water supply and sewerage systems of the RF', accounting for the water delivered and wastewater discharged will be undertaken by the consumer.

In order to account for water consumption and water disposal volumes the consumer shall arrange to have a metering station at each connection (customer service pipe) of its assets to the central water supply system. The consumer is responsible for the technical state of the metering station and water meters.

Accounting for drinking water consumption or the amount of wastewater discharged, setting up and keeping the required accounting documentation (logs, diagrams, diskettes and so on), making calculations and keeping reports to determine the volume of drinking water (discharged wastewater) delivered in an accounting period will be undertaken by the consumer.

The taking of readings from measuring devices used in settlements with the Vodokanal and submission of data about the volume of drinking water (or discharged wastewater) delivered to the Vodokanal will be undertaken by the consumer within terms defined in the contract.

The Vodokanal will ensure that the readings taking and data submitted about the volume of drinking water (discharged wastewater) are correct. If during an inspection discrepancies are detected between the readings from the measuring devices and data submitted by the consumer the Vodokanal will undertake a recalculation of the volume of drinking water (discharged wastewater) delivered during the period since the previous inspection until the discrepancy was detected from the readings given by the measuring devices.

A consumer or an organisation operating a metering station under contract with the Vodokanal will provide free access to Vodokanal representatives to the metering unit for inspection of the measuring devices and must submit documentation at the representative's request in order to confirm that the calculation for the volume of drinking water (discharged wastewater) delivered is correct.

e. Issues of rationing and quality control of wastewater discharged by consumers to the central wastewater disposal system

In accordance with the Resolution of the RF government of 31 December 1995 No 1310 'About payment collection for wastewater and pollutant disposal into sewerage systems in population centres', executive bodies of the RF's constituent parts determine the procedure for collecting payments from consumers for wastewater and pollutant discharges into the sewerage systems in population centres, taking into consideration any economic impact damage incurred to the sewerage systems and environment, including any excess quantities of wastewater and pollutants discharged.

Water disposal (discharge) standards for wastewater composition will be set for consumers by local authorities or by the Vodokanal on the instruction of a local authority (Clause 61 of the 'Rules of use for the municipal water supply and sewerage systems of the RF'), taking into consideration the following:

- standards of maximum allowable discharge for waste water and pollutants into water bodies, which are approved for the Vodokanal by the environmental authorities
- design parameters for wastewater treatment at treatment facilities prior to their discharge into water bodies

- the engineering and process capabilities of the wastewater facilities at treatment works to treat wastewaters from specific pollutants
- the need to protect networks and facilities at treatment works from destruction

For the period necessary for a consumer to implement measures for the rational use of drinking water and wastewater reduction approved by the Vodokanal, the Vodokanal can set temporary conditions for wastewater receipt as specified in the contract.

It is forbidden to discharge to treatment works wastewater containing substances that can (Clause 63 of the 'Rules of use of municipal water supply and sewerage systems of the RF'):

- litter pipelines, wells, screens or be deposited on the walls of pipelines, wells and other facilities of treatment works
- have a destructive effect on the materials of pipelines, equipment and other facilities of treatment works
- form explosive and toxic gas-vapour mixtures in wastewater networks and facilities
- prevent bacterial treatment of wastewater

It is forbidden to discharge substances and their transformation products for which maximum allowable concentrations are not established and for which there are no analytical control methods, and also any substances whose presence can lead to the formation of compounds with unestablished effects.

The consumer will provide laboratory control and observe the specified requirements and standards for the composition of wastewater discharged into sanitary works. Control of the consumer's compliance with wastewater disposal standards for wastewater composition will be undertaken by the Vodokanal through analysis of wastewater samples taken in the consumer's control wastewater wells.

A consumer is entitled to take part in wastewater sampling undertaken by the Vodokanal. Wastewater sampling will be confirmed by a certificate signed by the Vodokanal and consumer representatives. Analysis of control wastewater samples will be undertaken by the Vodokanal's laboratory or at his discretion in another laboratory accredited to undertake such work.

During the taking of control samples by the Vodokanal, a representative of the consumer can simultaneously take a sample and undertake its analysis in an independent accredited laboratory on their own account. If the results of the analysis of the samples are different and the Vodokanal laboratory is not accredited, then the true value will be acknowledged as the results obtained by an independent accredited laboratory.

If both laboratories are accredited and the Vodokanal insists that its laboratory results are correct, then a customer is entitled, in order to protect their interests, to address the accreditation body, which will check the analysis results of the laboratories and will take final decision on the matter under consideration.

f. Issues of Vodokanal settlements with consumers for water supplied and wastewater accepted

The Rules of use for municipal water supply and sewerage systems of the RF envisage that settlements with consumers for water supplied and wastewater accepted will be undertaken on the basis of data from measuring devices and laboratory control data regarding the quality of the wastewater discharge from the consumer (Clause 69 of the Rules).

In accordance with the Resolution of the Supreme Soviet of the Russian Federation of 1 April, 1993 No 4725-1 'About measures on settlement optimisation for production and services of municipal power and water supply companies and sanitary works' and the Resolution of the RF Government of 31 December 1995 No 1310 'About payment collection for wastewater and pollutant disposal into sewerage systems in population centres', settlements for water supplied and wastewater discharged will be undertaken on the basis of billing documents submitted by the Vodokanal to the bank without consumer agreement. The unsolved problem at present is the absence of regulatory documents about advance payments for services rendered by the Vodokanal.

g. Issues of the limitation or termination of water supply and wastewater disposal

The Vodokanal is entitled to limit or terminate a water supply and wastewater disposal to consumers without preliminary notification in the following cases:

- where the power supply to water and sewerage assets is cut off

• accidents as a result of Acts of God and emergencies
• a need to increase the water supply to fire access points

The Vodokanal is entitled to limit or terminate water supply and wastewater disposal to consumers with prior notification to the consumer, local authorities, local sanitary and epidemiological inspection bodies and also the territorial department of the fire fighting service of the RF in the following cases:
• an abrupt deterioration in quality in the water supply source
• receipt of an instruction or decision from the local sanitary and epidemiological inspection bodies
• unauthorised use of central water supply systems and/or wastewater disposal systems
• the entrance of unauthorised discharges of wastewater and pollutants into the central wastewater disposal system, leading to damage to the system or to an accident
• actions to deal with the consequences of an accident within the central water supply systems and/or wastewater disposal systems
• an emergency or unsatisfactory condition within the water supply and/or wastewater networks of a consumer
• when handling works on a new consumer connection within terms approved by the above authorities
• scheduled maintenance of networks and facilities of the central water supply and wastewater disposal systems
• a failure to pay by a consumer for two or three accounting periods for their water supply and wastewater disposal services within the terms stipulated by the contract

In case of a limitation or termination of the water supply for failure to pay, the Vodokanal shall first notify the consumer seven days before the limitation begins. If within seven days of receiving notification the consumer does not pay their debt, the limitation will be introduced. If within 10 days of the limitation being introduced, the consumer does not pay their debt then the water supply and wastewater discharge can be completely terminated.

2.3.6 Other control measures

Vodokanals are obliged to provide services in accordance with the requirements of federal, regional and local laws, regulations and other regulatory documents. Their observance is mandatory whatever the form of Vodokanal incorporation.

a. Licence

The Vodokanal licence for water abstraction usually covers the following aspects: the expected volumes of water consumption by domestic and other consumers, water consumption volumes for parts of the Vodokanal and water losses. Where water consumption is in excess of the volumes set by the licence, the Vodokanal can have penalties imposed.

b. Water Code, Rules for use of municipal water supply systems and Rules for provision of public utility services

Verification of the Vodokanal's compliance with the requirements of regulatory documents relating to water supply and wastewater disposal services, including the Water Code (Water Code of the RF (Part 2, Article 6)), Rules for use of municipal water supply and sewerage systems of the Russian Federation (Resolution of the RF Government No. 167 of February 12, 1999 About approval of Rules for use of municipal water supply and sewerage systems of the Russian Federation) and Rules for provision of public utility services (Resolution of the RF Government No. 1099 of September 26, 1994 About approval of Rules for provision of public utility services and Rules for hard and liquid waste disposal in the Russian Federation), is usually undertaken by the local administration.

c. Environmental safety

Verification of the Vodokanal's compliance with the requirements of the Resolution of the RF government of 31 March 2003 No 177 'About the organisation and implementation of state environmental monitoring (state ecological monitoring)' is carried out by local environmental committees or similar authorities and regional/ municipal environmental committees and inspections.

d. Health care
Verification of the Vodokanal's compliance with the requirements of the Federal Law of 30 March 1999 No 52- FZ 'About sanitary and epidemiologic protection of the population', is carried out by regional Centers of State Sanitary and Epidemiological Inspection and regional or municipal departments of State Sanitary and Epidemiological Inspection.

e. Safety
The Federal Agency for Ecological, Engineering and Nuclear Supervision (Rostekhnadzor) carries out control and supervision of compliance for requirements for asset (facility) registration in the state register of hazardous facilities as well as compliance with industrial safety requirements during the design, construction, operation, temporary closing and liquidation of hazardous facilities, including requirements to industrial safety declaration issue for hazardous facilities. In case of Vodokanals this function is mainly referred to objects where chlorine is used for preliminary water treatment and disinfection.

2.4 Tariff regulation principles and procedures
2.4.1 Analysis of current legislation
The main principles and procedures relating to tariff regulation (as of 01 March 2006) are envisaged by the following federal laws and regulatory documents:
• Tax Code of the RF
• Federal law of 30.12.2004 No. 210- FZ 'About the tariff regulatory framework of municipal housing enterprises'
• Resolution of the RF government of 17.02.2004 'About approval of the basis for price formation in the municipal housing economy'
• Resolution of the RF government of 07.03.1995 No. 239 'About measures on normalisation of state regulation of prices (tariffs)'
• Resolution of the RF government of 05.11.2003 No. 674 'About the procedure for considering discrepancies between executive authorities of the RF constituents in the state regulation of tariffs by organisations exercising regulated activities and consumers'
• Resolution of the RF Government of February 13, 2006 No. 83 'About approval of the Rules for determination and provision of engineering conditions for connection of capital construction object to engineering utilities' and the 'Rules for connection of capital construction assets to engineering utilities'
• Order of Federal Tariff Agency (FTA) of 29.12.2005 No. 853- E/4 'About approval of maximum allowable change indices for schedule rates for goods and services of municipal housing enterprises'
• Order of FTA of December 7, 2004 No. 236- E 'About approval of the approval procedure for decisions of executive authorities of the RF constituents in the area of state regulation of tariffs about setting of tariffs exceeding the limit level established by the federal executive authority in the area of state regulation of tariffs'

2.4.2 Main definitions used in tariff regulation
Production programme: a programme of Vodokanal activities relating to provision of water supply and wastewater treatment and disposal services, which includes measures for the refurbishment of the municipal infrastructure system operated by such a Vodokanal.

The production programme is developed by the Vodokanal and approved by its regulatory body (the executive body of that RF constituent, if consumers of this Vodokanal's services live in several Municipalities and none of them uses more than 80% of its services, and by the local authority in other cases).

Integrated development programme of municipal infrastructure: a programme of construction and/or reconstruction of municipal infrastructure systems undertaken by a Municipality, which provides the following:
• development of these systems in accordance with the requirements of residential and industrial construction
• improvement in the quality of services provided to consumers
• improvement in the ecological situation in the territory of a Municipality

An integrated development programme for municipal infrastructure has to be developed by the local authority and approved by a delegate body of the Municipality. These programmes have to be developed and approved before 1 January 2010.

An investment programme is a programme of construction financing and/or municipal infrastructure reconstruction set out by the local authorities for the Vodokanal to implement the integrated development programme of municipal infrastructure.

The technical specification for developing the investment programme is prepared by the Vodokanal in accordance with the integrated development programme of municipal infrastructure and is approved by the head of local administration.

If an integrated development programme for municipal infrastructure is not available, the technical specification for investment programme development will be approved by the responsible department from the Municipality.

The investment programme is developed by the Vodokanal, reviewed and coordinated with the local authority and approved by a responsible body within the Municipality.

Tariffs for Vodokanal services are rates paid on the basis of services provided by the Vodokanal (without consideration of surcharges to tariffs for their services) and which are included within the price (tariff) for consumers.

Tariffs for Vodokanal services should secure the financial requirements of the Vodokanal for implementing its production programme.

Tariffs for Vodokanal services are calculated by the Vodokanal and set by a regulating body (an executive body of that part of the RF, if the consumers of these services live in several Municipalities and none of them consumes more than 80% of its services; otherwise they are set by the local authority).

A surcharge to the Vodokanal's services tariffs is a tariff set up for the Vodokanal as an extra charge to the consumers' payments to the Vodokanal for services provided; it is also used to finance the Vodokanal's investment programme.

The surcharge to tariffs for Vodokanal services must ensure implementation of the investment programme by the Vodokanal.

The surcharge to the Vodokanal's services tariffs must ensure that the Vodokanal's investment programme can be implemented by the Vodokanal, in coordination with the local authority and approved by a responsible department from the Municipality.

Tariffs for consumers are tariffs that include tariffs for Vodokanal services and those of other of public utility enterprises that provide water supply, wastewater disposal and wastewater treatment services (without taking into account consumer tariff surcharges).

The surcharge to the consumer tariffs is a tariff estimated during payment for the services provided by the Vodokanal and other public utilities, and is set in order to finance the investment programme of the Vodokanal and the other public utility enterprises; its total amount corresponds to the sum of the surcharges and tariffs for services provided by the Vodokanal and other public utilities implementing investment programmes to develop municipal infrastructure.

Tariffs and surcharges to tariffs for consumers are calculated by a regulatory authority that undertakes an analysis of consumer affordability.

The Vodokanal's tariff for connection to the municipal infrastructure system is a rate comprised of the charge for connecting new constructed or reconstructed real estate objects to utility networks. The rate is set for the Vodokanal and is used to finance its investment programme.

The connection charge imposed on connections to utility networks is a charge paid by entities undertaking the reconstruction of a building or facility if the reconstruction will lead to an increase in demand by that building or facility.

Consumer tariff affordability is the purchase and payment affordability for the consumers of a Vodokanal's goods and services with the surcharge taken into account.

Tariff affordability for consumers is set by:

- regulatory bodies of that part of the RF, for consumers of the services of a Vodokanal that are located across several Municipalities and do not use more than 80% of this Vodokanal's services
- regulatory bodies of the Municipality, for other Vodokanals

The set of criteria used to determine consumer tariff affordability is decided by the regulatory bodies of that part of the RF.

The financial needs of a Vodokanal are calculated amounts of money received for services provided, which are required to implement the production programme and/or investment programme of the Vodokanal.

Modification of objectives for the operating conditions of the Vodokanal, which include:

- changes to the Russian Federation's legislation
- an inflation rate exceeding the rate of inflation used in the calculations for tariff approval
- a shift in tariffs for fuel and power resources exceeding the amounts set in the regulatory documents of the Russian Federation
- modification of other conditions determined in accordance with an order established by the RF Government.

Limit indices are indices of maximum and minimum tariff changes for Vodokanal services, which have to be applied by the end of the current year, set on average by the individual parts of the RF and/or the Municipalities for the following business year.

Each year, ahead of the introduction into federal law of the federal budget for the next business year to the State Duma of the RF Federal Assembly, the Federal Tariff Agency (FTA) sets limit indices as an average for the various parts of the RF. Specific limit indices are set taking into consideration the investment programmes of the public utility companies approved by the representative bodies of their local authorities and also taking into account regional and other peculiarities. The FTA is only entitled to set indices for the maximum tariff shift for goods and services with tariff surcharges for goods and services taken into account.

The executive bodies of the Russian Federation, in accordance with the limit indices set by the FTA for their specific part of the Russian Federation, set limit indices for the Municipalities prior to adoption of the federal law regarding the federal budget for the next business year. This is undertaken on the basis of the amount of services provided by the Vodokanal, taking into account the approved investment programmes and local peculiarities.

If decisions about tariff setting for 2006 are taken by regulatory bodies before 1 January 2006 and these tariffs exceed the limit indices set by the FTA then the tariffs have to be revised downwards.

Provisions for the regulation and application of limit indices will be valid until 1 January 2009. The respective regulatory bodies will examine tariffs and their surcharges and evaluate their consumer affordability.

2.4.3 General principles of tariff regulation

The general principles of tariff and surcharge regulation are:

- striking a balance of interest between achieving the Vodokanal's consumer service goals, providing affordability for consumers and the effective functioning of the Vodokanals
- setting tariffs and surcharges to ensure that the financial requirements of the Vodokanals needed to implement their production programmes and investment programmes are met
- stimulating a reduction in the Vodokanal's operating expenses, increasing the economic efficiency of their services and encouraging use of energy-saving technologies
- creating the conditions needed to attract investment in order to develop and rebuild the municipal infrastructure systems
- fully compensating the Vodokanal for expenses related to implementing its production and investment programmes
- arranging the conditions for mandatory tariff changes for Vodokanal services
- ensuring access to information for consumers and other individuals regarding the formulation of tariffs and surcharges

2.4.4 Powers of the RF Government in the field of tariff and surcharge regulation for water supply and wastewater disposal services

The Government of the Russian Federation, in accordance with the Federal Law of 30 December 2004 No 210- FZ implements the following powers in the field of tariff and surcharge regulation:

The RF government approves:

- the basis for price formulation and the rules of tariff and surcharge regulation for water supply and wastewater disposal services
- the rules of tariff regulation for connections to municipal water supply and sewerage systems
- mandatory rules for the parties involved in public contracts to be concluded as part of the process of connecting to municipal water supply and sewerage systems

The RF government sets:

• the federal executive authorities' approval procedure for the Unified Reporting System for public utility companies and the period for submission of reports
• the monitoring procedure for implementation of production and investment programmes by the public utilities
• other guidelines and recommendations relating to the preparation of technical assignments for the development of investment programmes to be undertaken by public utilities
• the coordination procedure for the executive authorities of the various parts of the RF in regulating the tariffs for public utilities' goods and services and for local authorities regulating the tariffs and surcharges of public utilities
• the settlement procedure for disputes arising between executive authorities of the parts of the RF regulating tariffs for public utilities' goods and services and the local authorities regulating tariffs and surcharges of public utility companies and for public utilities in relation to the setting of tariffs and surcharges

2.4.5 The procedure for determining the operating environment of public utilities, whose modification impacts on the costs of their goods and services

Composition of the Vodokanals' production programme.

The production programme includes:

• justification for the predicted volume and quality of services provided by a Vodokanal in accordance with the requirements of its technical regulations, environmental standards and existing production capacities of the Vodokanal
• an action plan for efficiency upgrades to Vodokanal activities, which envisages improvement in the quality of services and, if required, the reconstruction of the municipal infrastructure operated by the Vodokanal.

In the case where, during a period when the tariff is valid, a Vodokanal reduces costs (in comparison with the values envisaged in the production programme) because of an increase in the efficiency of its activities, the saved funds will remain at the Vodokanal's disposition. In this case the current tariffs cannot be reduced as the savings were made while applying the tariffs.

2.4.6 Methods of tariff regulation

The method of tariff regulation will be determined by the regulatory bodies (the executive bodies of the individual parts of the RF if services are provided by one Vodokanal to consumers located in several Municipalities and none of them uses more than 80% of this Vodokanal's services; otherwise by the local authority).

Methods of tariff regulation for water supply and water disposal services include the following:

• the setting of fixed tariffs for Vodokanal services for the next tariff period on the basis of the cost of goods and services in the previous tariff period, taking into account measures to increase efficiency included in the Vodokanal's production programme or envisaged improvements in the quality of produced goods (services provided) and any reconstruction measures for the municipal infrastructure system operated by the Vodokanal
• the setting of limit tariffs for Vodokanal goods and services is determined on the basis of a cost behaviour analysis of the organisation's previous activities and an analysis of similar activities of other public utilities
• indexation of tariffs set for Vodokanal goods and services in cases where the operating environment is modified, impacting on the cost of services provided, which are specified by the law

During the process of regulation, different combinations of tariff regulation methods can be applied to Vodokanal services.

2.4.7 Tariff regulation procedures

Tariff regulation for Vodokanal services includes:

• coordination of the Vodokanal's production programme with a regulatory body
• setting of tariffs for Vodokanal services to provide sufficient cash inflows from its services to implement its production programme, taking into account affordability for Vodokanal consumers and the set limit indices

The Vodokanal develops a production programme for the next tariff period for its services, determines the financial requirements for implementing the programme and makes a preliminary calculation of tariffs for its own goods and services sufficient to ensure its requirements are met.

The production programme, the calculation of financial requirements and the various tariffs will be sent by the Vodokanal to a regulatory body within three months of the end of the current valid tariff period.

A regulatory body (the executive bodies of the constituent parts of the RF if services are provided by a Vodokanal to consumers located in several Municipalities and none of them consumes more than 80% of its services, otherwise a local authority) shall check the justification for the production programme, tariff calculations will determine the affordability of Vodokanal services.

When the production programme is unjustified or does not comply with the Vodokanal's production activity indicators determined by monitoring, the regulatory body will be entitled to recommend changes to the production programme and to send back this programme and its calculations to eliminate any discrepancies that may be found

When the calculated tariffs for goods and services do not meet the requirements of the Vodokanal to finance the implementation of its production programme, the regulatory body will be entitled to undertake independent calculations of these tariffs.

If the calculated tariff makes the Vodokanal's services to its consumers unaffordable, the Vodokanal, in agreement with its regulator, will make changes to its production programme and undertake a recalculation of its tariffs for services.

After completion of the checks on the justification for the calculation of financial requirements, submitted by a Vodokanal and required to implement its production programme, and the proposal of tariffs complying with these needs, the regulatory body approves the production programme and set tariffs for the Vodokanal's services corresponding to this programme.

Tariffs for Vodokanal services for the next tariff period will be set not less than one calendar month before the end of the current valid tariff period.

2.4.8 Procedure for review and approval of the investment programme, surcharges to tariffs and connection charges

The local authority develops the integrated development programme of municipal infrastructure, which is approved by the responsible body within the Municipality.

The head of the local administration of the Municipality, in accordance with the integrated development programme of municipal infrastructure, approves the technical specification for development of the investment programme.

The Vodokanal develops a draft investment programme and calculates the financial requirements for its implementation, which are submitted to a regulatory body within the Municipality.

The regulatory body within the Municipality checks the draft investment programme and the conditions of the approved technical specification for the formulation of its financial needs calculation and justification check for implementing the programme; after this it will prepare suggestions about:

- the size of the surcharge to tariffs for Vodokanal services
- the size of the charge for connection to the municipal infrastructure system

The charge for connection to the municipal infrastructure system will be set per unit of declared load for buildings and facilities under construction or per unit of declared load for buildings and facilities under reconstruction.

The Municipality regulatory body will undertake an affordability analysis of the suggested surcharge to tariffs for Vodokanal services to consumers and the charge for connection to the municipal infrastructure system.

If the Municipality regulatory body decides that the surcharge to tariffs and connection charge are unaffordable for Vodokanal customers, then that body can:

- prepare suggestions for changes to the conditions of the technical specification, on the basis of which the investment programme was developed
- prepare suggestions to ensure the financial requirements of the Vodokanal are partially met from local budget funds.

After taking a decision about the affordability of the surcharge to tariffs and the connection charge for Vodokanal services, the Municipality regulatory body will send the draft investment programme and the calculations of surcharge to tariffs and connection charge to the responsible department within the Municipality.

The responsible department of the Municipality will review and approve the investment programme and set the surcharge to the consumer tariffs.

The Municipality regulatory body will set tariffs for Vodokanal services, charges for connection to the municipal infrastructure systems and also the Vodokanal connection charge.

When all of the specified tariffs and surcharges are set, the local authorities conclude a contract with the public utility company in order to develop the municipal infrastructure systems; the contract determines the conditions for implementing this company's investment.

The Vodokanal's investment programme and/or terms of its implementation can be revised by the responsible department within the Municipality at the request of the Vodokanal, the Municipality regulatory body or by its own initiative:

• if, as a result of monitoring the implementation of the investment programme it is found that the profitability of the operations is much higher or lower than the break-even level calculated during approval of the investment programme

• in the case of a modification to the Vodokanal's operational environment that impacts the cost of its services, and when it is impossible to review the surcharge to the Vodokanal's tariffs and/or connection charge.

The size of the connection charge is determined by the amount of declared load (the increase in throughput of the reconstructed asset) for the asset being connected to the municipal water supply and wastewater systems.

2.4.9 Validity period of tariffs and surcharges

The valid period for tariffs for Vodokanal services cannot be less than one year and it must agree with the implementation period for its production programme.

The valid period for the Vodokanal's connection charges and tariff surcharges for Vodokanal services will not be less than three years in each case and they must agree with the implementation period for its production programmes (the stages of the investment programmes).

During the valid period for the surcharges to consumer tariffs, on the decision of the responsible department within the Municipality, a modification (increase or decrease) to consumer surcharges can be undertaken (at most once a year), relating to approval of the new investment programmes and/or in connection with the conclusion of the implementation of an investment programme.

2.4.10 Pre-schedule review of tariffs and surcharges

The grounds for a pre-schedule review of tariff regulation for Vodokanal services are:

• modification to the objectives of the Vodokanal's operational environment that impact on the costs of its services

• violation by the Vodokanal of its approved production programme detected as a result of monitoring this programme

• an instruction from the Federal Anti-monopoly Agency (FAA) or Federal Tariff Agency (FTA)

• an instruction from a regulatory body of the constituent part of the RF to a Municipality regulatory body

• a judgment, which has come into effect.

The ground for a pre-schedule review of tariffs for Vodokanal services and the connection is a modification of the objective of Vodokanal's operational environment that impacts on the costs of its services.

The grounds for a pre-schedule review of the surcharge to consumer tariffs and the connection charge are the following:

• a change in the Vodokanal's investment programme

• a pre-schedule review of the surcharge to tariffs for Vodokanal services and/or the connection charge

In case of a modification to the objectives for the Vodokanal's operational environment and a decision taken by a regulatory body regarding the affordability to consumers of the tariffs for

Vodokanal services, then the tariffs will be subject to pre-schedule review.

In case of a modification to the objectives for the Vodokanal's operational environment and a decision by a regulatory body that the tariffs for Vodokanal services cannot be afforded by consumers, then the production programme will be subject to a pre-schedule review without a tariff review.

A pre-schedule review of tariffs and surcharges can be undertaken at most once in a year, excluding a pre-schedule review of tariffs and surcharges on the instructions of the FAA and FTA or on the basis of a judgment coming into effect.

3. Analysis of the municipal water supply and sewerage systems in Russia and the policy of the Russian government on their reconstruction

3.1 General information

An analysis of the capacity and technical condition of the water supply and sewerage systems is undertaken by using statistical reporting data for 2002 to 2004 and is given by region and by the large administrative centres of the constituent parts of the RF.

3.2 Engineering performance condition of the municipal water supply and sewerage systems in Russia

3.2.1 Water intakes, waterworks and wastewater treatment facilities and pumping stations

Water supply

Statistical reporting data for 2004 in Russia show:

- the capacity of first lift water main pumping stations was 114.6 million cubic metres per day or about 800 litres per day per capita on average
- water treatment works' capacity (that is, the production capacity and supply of drinking water to the network) was 88 million cubic metres per day or 615 litres per day per resident of Russia on average.

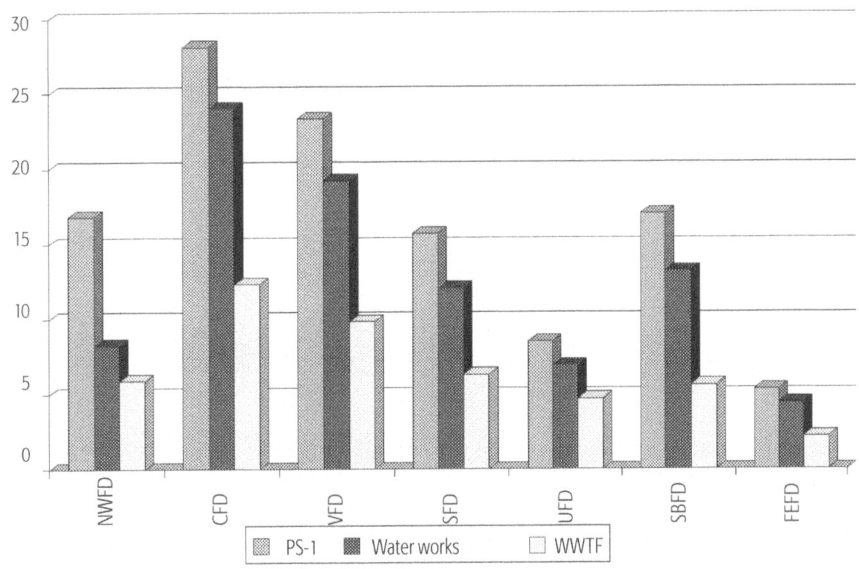

Figure 3.2.1(a). Works capacity by federal district at the beginning of 2005 (in Mm³ per year)

Figures 3.2.1(a) and (b) show a set of data about the capacity of first lift pumping stations (PS-1), water treatment works and wastewater treatment works (WWTWs) as of the beginning of 2005.

As can be seen, the output from specific water treatment works in Russia's federal districts does not vary greatly.

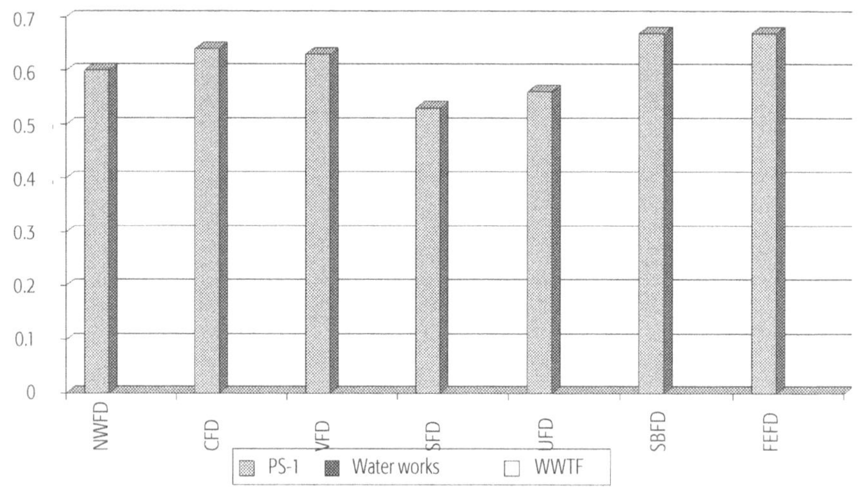

Figure 3.2.1(b). Specific output of works by federal district at the beginning of 2005 (in m^3/person/day)

Sewerage

In 2004 municipal sewerage system received 15.1 billion cubic metres of effluent or 41.4 million cubic metres per day. The volume of wastewater disposed of to the municipal sewerage was 290 litres per capita. The significant difference between the volume of water supplied and wastewater disposal is explained by the fact that in rural population centres there are no centralised sewerage systems.

At the same time 12.5 billion cubic metres or 83% of the effluent entering the municipal sewer system underwent complete biological treatment. In all, 14% of the effluent was subject to advanced treatment following biological treatment.

The total volume of wastewater discharged into water bodies included:

- effluents treated to the standard quality, which made up 25% of the wastewater volume received by the municipal wastewater systems
- insufficiently treated effluents, which made up 46% of the wastewater volume received by municipal waste water systems
- untreated effluents, which made up 29% of the wastewater volume received by the municipal wastewater systems

3.2.2 Water supply and sewerage networks

According to statistical data reports, as of the beginning of 2005 there were the following utility networks in Russia:

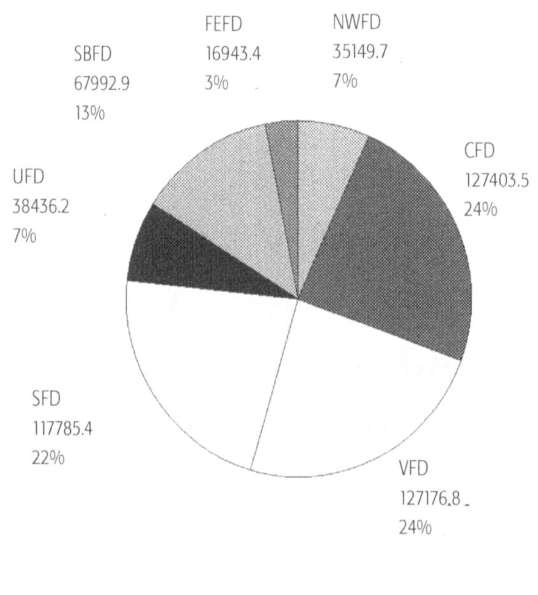

Figure 3.2.2(a). Length of water supply networks by RF federal district at the beginning of 2005 (in km)

- 132,000km of water conduits of which 30,000km or 23%, needs to be replaced
- 336,000km of mains water supply networks of which 127,000km, or 38%, needs to be replaced;
- 64,000km of sub-main water supply networks
- 45,000km of sewage collectors, of which 9000km, or 20%, needs to be replaced
- 77,000km of mains sewerage networks of which 25,000km, or 33%, needs to be replaced
- 54,000km of sub-main and building sewerage networks

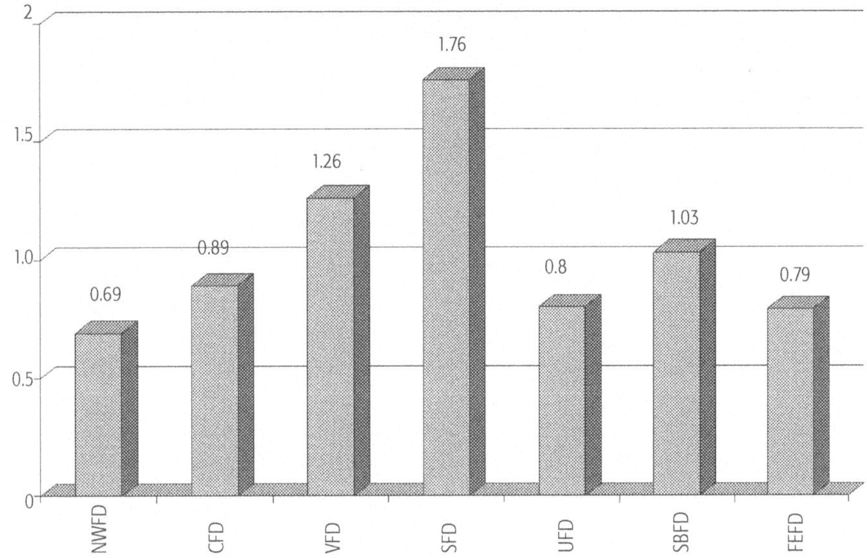

Figure 3.2.2(b). Requirements for replacement of worn water supply network by federal district (km per resident).

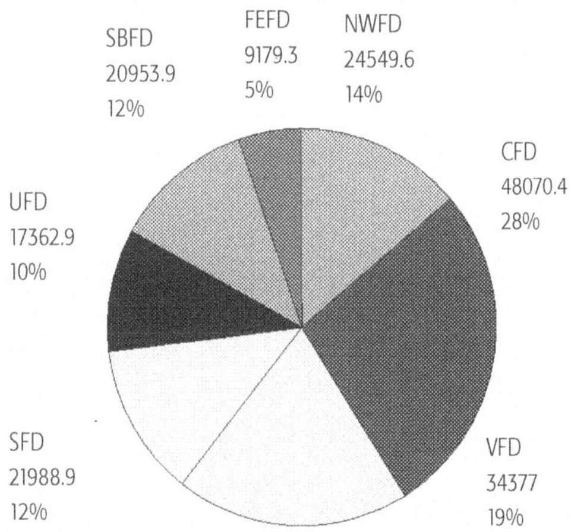

Figure 3.2.2(c). Length of sewerage networks by RF federal district at the beginning of 2005 (in km).

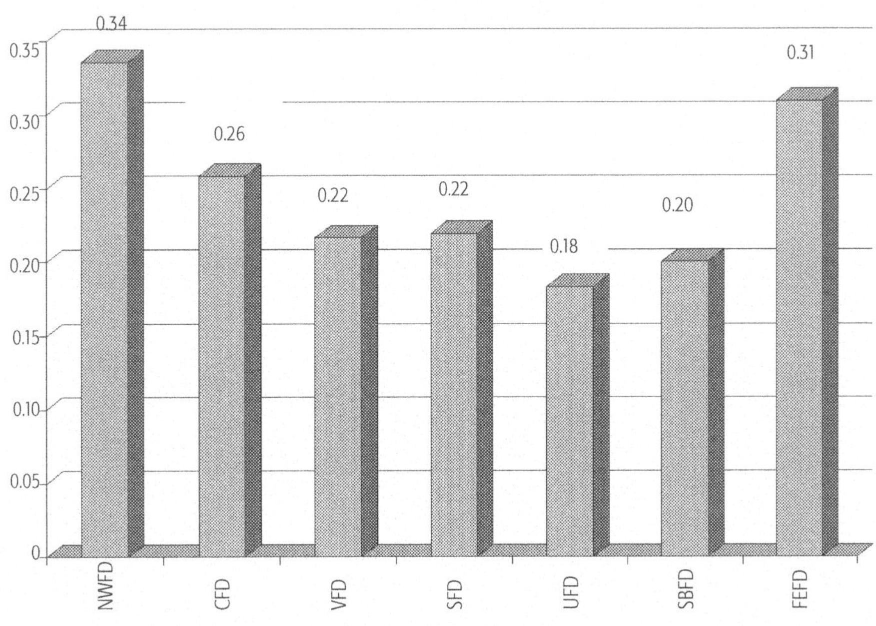

Figure 3.2.2(d). Requirements for replacement of worn sewerage networks by federal district (km per resident).

- 605,000 mains water pumps

The water supply networks are distributed among federal districts as shown in Figure 3.2.2(a). (Note: NWFD – North West Federal District; CFD – Central Federal District; SFD – Southern Federal District; VFD – Volga Federal District; UFD – Uralian Federal District; SBFD –Siberian Federal District; FEFD – Far Eastern Federal District.)

As seen in Figure 3.2.2(b), the residents of the Southern and Volga federal districts bear the biggest investment burden, twice as large as that of residents of the North-West, Uralian or Far East federal districts.

As seen in Figures 3.2.2(c) and (d), the investment requirement for replacing the sewerage network on a 'per resident' basis in the Volga and Southern federal districts are essentially lower in comparison with that facing the residents of the North-West and Far East federal districts.

3.2.3 Water losses and incident rates

It can be seen from Figures 3.2.3 (a) and (b) that the amount of losses and unaccounted-for water in the RF increases by 5% annually on average. If this trend does not stop, by 2020 the water losses will exceed 30% of the water supply going into the network. Nevertheless, this index cannot be considered reliable as in the majority of Russian cities less than 50% of the volume of water sold to consumers is metered.

As seen in Figure 3.2.3(c), the percentage of losses and unaccounted-for water varies considerably (from 8% to 40%). As mentioned above, this index is not always reflective of the actual situation because there is insufficient measurement of consumption.

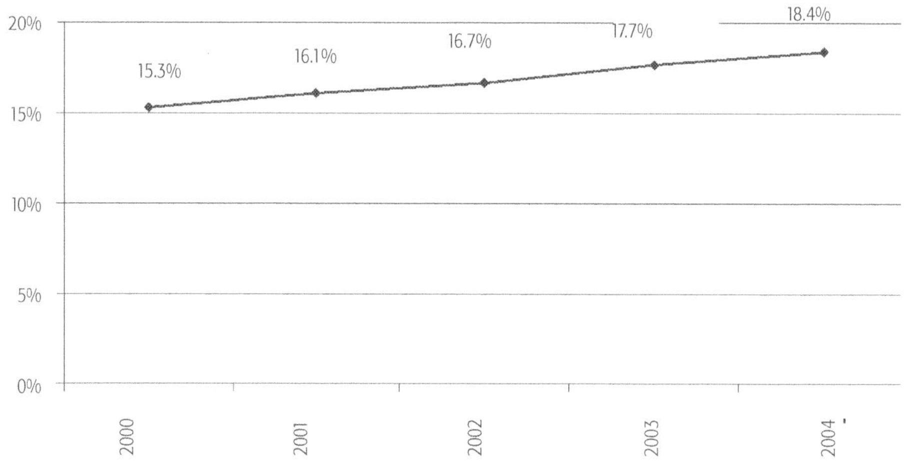

Figure 3.2.3(a). Losses and unaccounted-for water consumption in the RF in 2000-2004 (in % of water supply volume to water distribution mains).

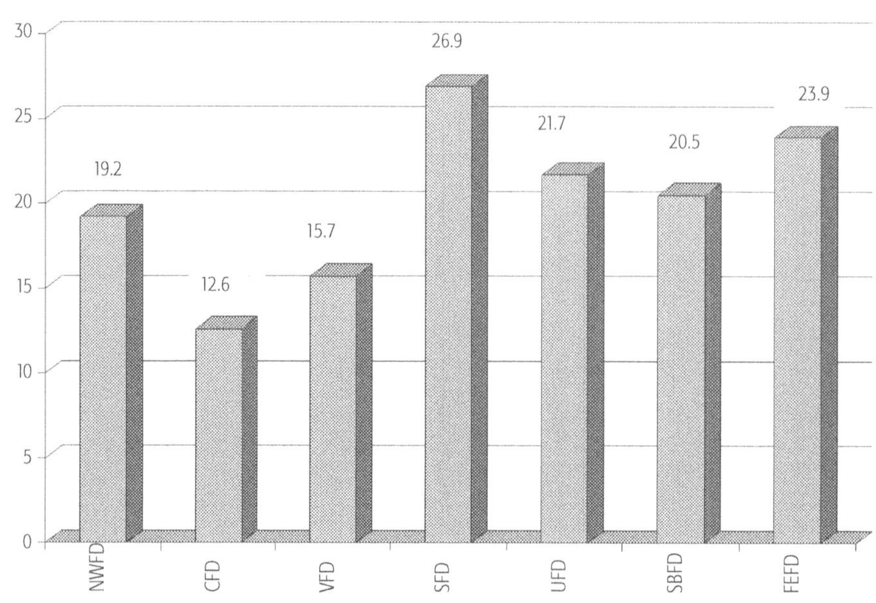

Figure 3.2.3(b). Losses and unaccounted-for water consumption by federal district in 2004 (in % of water supply volume in water distribution mains).

Analysis of statistical data reports that relate to this incident rate does not ensure an objective result, as the definition of an 'incident' is interpreted in different ways by different Vodokanals: in one case the term covers situations relating to any termination of the water supply to consumers, in another, it might cover situations where a termination of the water supply to consumers exceeded a set time (as a rule, more than eight hours). That it is why it is necessary to be fairly careful with this index.

When this index is compared with the water losses and unaccounted-for consumption index there is no relationship (Figure 3.2.(d)).

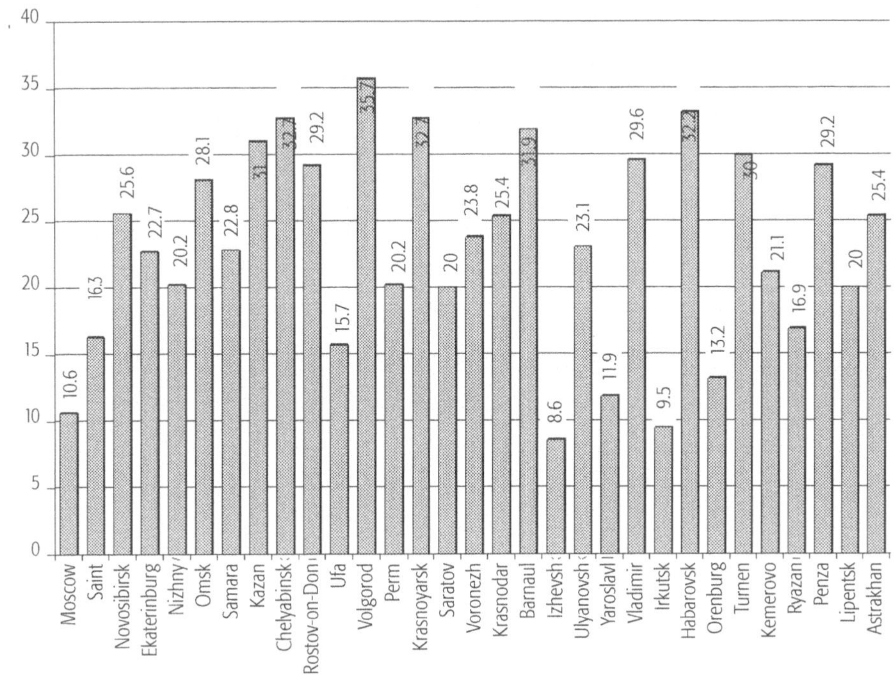

Figure 3.2.3(c). Losses and unaccounted-for water consumption in cities of Russia with population size more than 500,000 in 2004 (in % of water supply volume to water distribution mains).

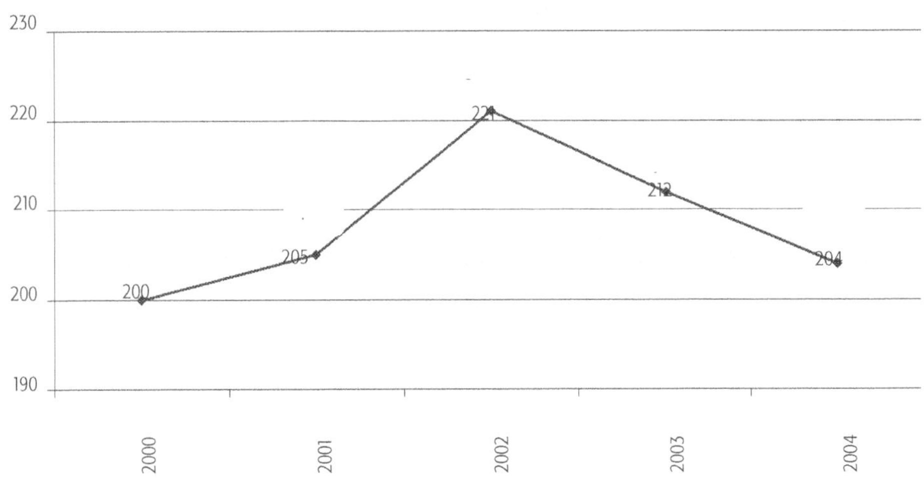

Figure 3.2.3(d). Number of incidents in the RF in 2000-2004 (thousands of incidents per year). per resident).

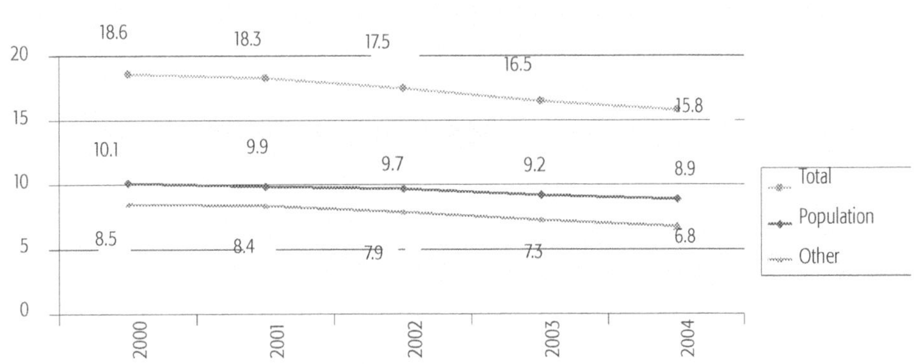

Figure 3.3.1. Volume of water supplied to consumers in 2000-2004 (Mm³).

3.3 Economic indices of the water supply and wastewater sector in Russia

3.3.1 Volume of services rendered to consumers

As seen from Figure 3.3.1, the volume of services rendered to consumers decreases in real terms by an average of 4% annually both for water supply and wastewater.

3.3.2 Cost of services rendered to consumers

Despite the decrease in real terms of the volume of sales, the cost of the services rendered is growing due to tariff increases.

Figure 3.3.2(a) is a history of the changes in the cost of water supply and wastewater disposal services in Russian between 2001 and 2004.

Figure 3.3.2b) shows a history of payment for accommodation and public utility tariff collection from the population between 2001 and 2004.

As can be seen from the diagram, the collection of payments from the public is increasing (despite the growth in public utility tariffs outstripping inflation). The growth in collection from the public is connected, primarily, with a rise in living standards.

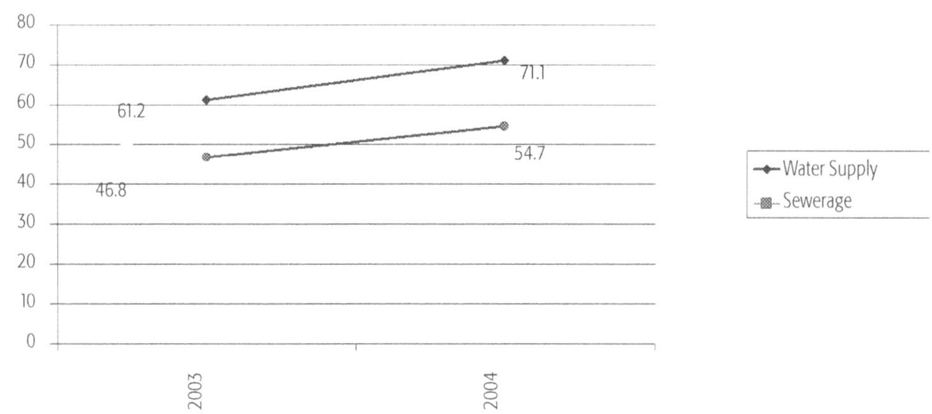

Figure 3.3.2(a). Changes in cost of services rendered to consumers in 2003-2004 (billion roubles).

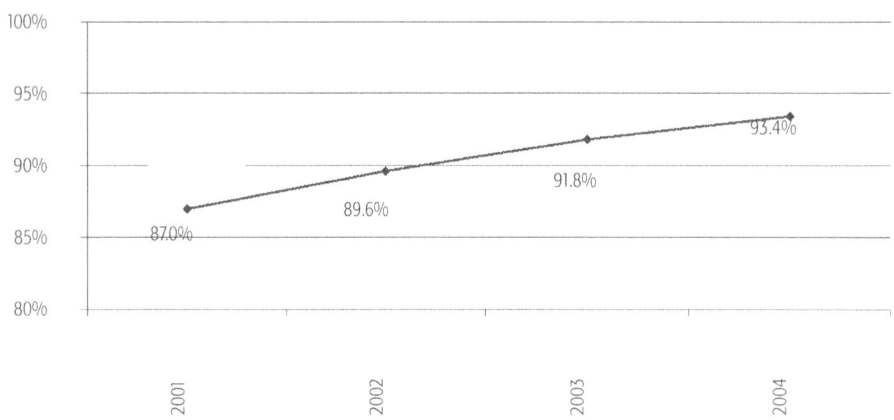

Figure 3.3.2(b). Accommodation payment and public utility tariff collection, 2001-2004 (as %).

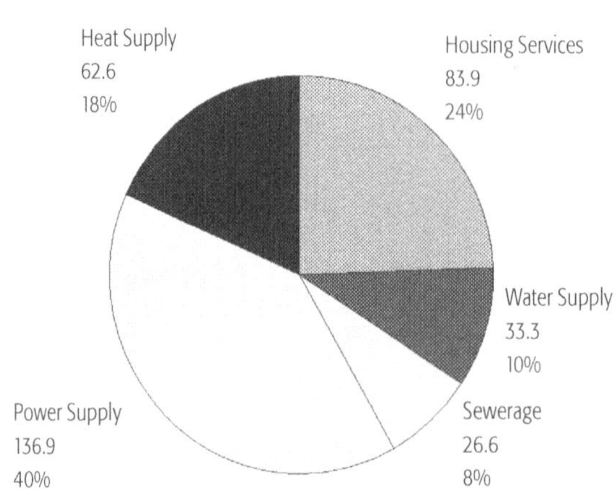

Figure 3.3.2(c). Structure of payments for accomodation and public utility services in 2004 (billion roubles).

Figure 3.3.2(c) is a diagram representing the structure of customer payments in terms of the total volume of public utility services provided in 2004.

3.3.3 Time history of average tariffs for water supply and sewerage services
Figure 3.3.3 shows the average tariffs for water supply services for 2002-2004.

3.3.4 Fixed asset costs of municipal water supply and sewerage systems in the RF
As can be seen from Figure 3.3.4, the cost of the fixed assets grows constantly with the decrease

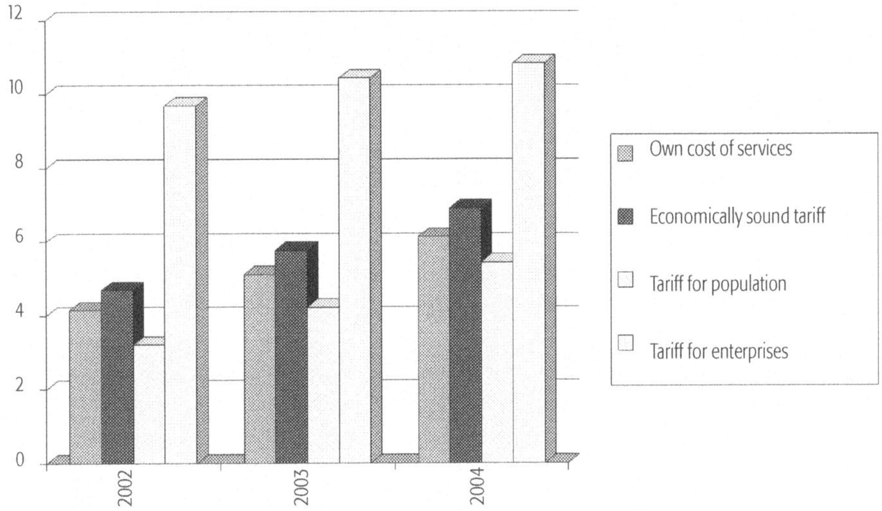

Figure 3.3.3 Average tariffs in the RF for water supply services, 2002-2004 (roubles/m³).

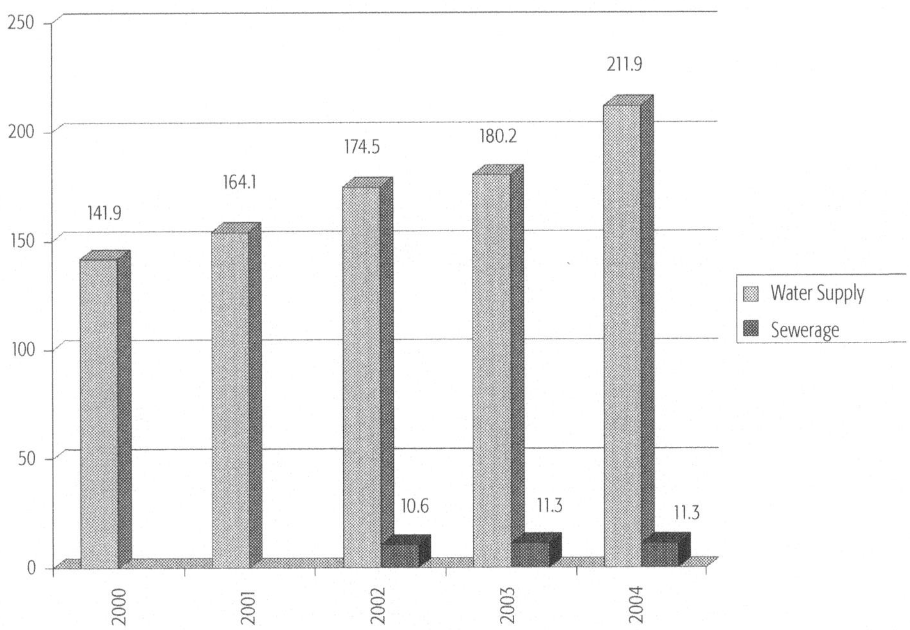

Figure 3.3.4. Fixed asset costs of municipal water supply and sewerage systems in the RF.

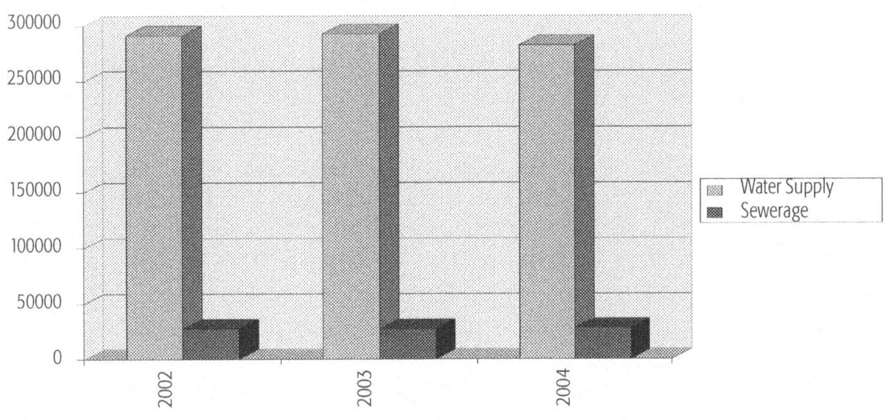

Figure 3.3.5. Number of employees servicing municipal water supply and sewerage systems.

in services sale volume, which characterises the decrease in the return on the fixed assets.

3.3.5 Personnel

As can be seen from Figure 3.3.5, the number of employees within the municipal water supply and sewerage systems is not effectively increasing; the increase in the cost of the fixed assets provides an increase in the efficiency of the maintenance and at the same time the decrease in the volume of service sales reduces the sector's labour efficiency.

3.4 Housing programme

In 2003 to 2004, within the framework of implementing the sub-programme for reform and reconstruction of the municipal housing economy of the Russian Federation, state support was given for the construction or reconstruction of 240 municipal infrastructure assets in 63 constituent parts of the RF.

In all, 8,975km of water supply and sewerage networks, heating boiler plants with a total capacity of 22,220Gcal/h, wastewater treatment facilities and water supply and wastewater disposal assets with a total capacity of 4,938,000m^3/day were rebuilt.

The wear and obsolescence found in municipal infrastructure assets are due to the low quality of public utility services, which does not meet consumer requirements. In all, 60% of municipal infrastructure assets on average are worn out. In addition, 55% of the boilers in the Russian Federation (as of 1 January 2005) were worn out, as were 65% of the water supply networks, 63%, of the sewerage networks, 63% of the heat supply networks, 58% of the power supply networks, 65% of the water supply pumping stations, 57% of the sewerage pumping stations, 54% of the water treatment works, 56% of sewage treatment facilities and 57% of transformer substations.

There is a general lack of compliance with the requirement to inject the actual amount of investment that is needed into the reconstruction of municipal infrastructure assets to enable them to meet their minimum standards. Planned maintenance of water supply networks and equipment for municipal power supply systems has effectively been completely replaced by emergency works. This has led to a decrease in the reliability of municipal infrastructure assets.

The inefficient use of natural resources has led to high water, heat and power losses during resource production and transportation to consumers.

As a result of the wear-out found in the municipal infrastructure, the total losses in the heat supply networks are 30% of the heat produced, which is equal to the consumption of 65 to 80 billion units of equivalent fuel per year. Heat media leakage, resulting from pipeline corrosion, has resulted in heat losses of 10% to 15%. The dilapidated condition of the heat and power supply networks is the reason for the cuts to the heat supply in residential blocks in winter.

Losses and unaccounted-for water in the water supply systems in a number of cities has reached 60% of the water put into the mains. A reduction in the standard of drinking water quality supplied to the public within the Russian Federation is one result of this situation. More than 40% of water mains with surface water intakes, which supply 68% of the demand in towns and urban villages, and about 65% of the water mains in rural areas do not have the necessary water treatment works and do not undertake full disinfection and water treatment that complies with the requirements for drinking water quality.

Shortfalls in wastewater treatment have created environmental pollution. In all, 60% of wastewater treatment works are overloaded, 38% have been operating for 25 to 30 years or more and require urgent reconstruction. The under-capacity at wastewater treatment works has now reached about nine billion cubic metres per day. As a result, untreated or insufficiently treated wastewaters are discharged into open water bodies, which aggravates existing ecological problems and decreases the recreational attractions of the water bodies.

Nevertheless, the cost of public utility services has increased significantly in recent years. The 'cost is no object' approach current in most of the tariff formulations for heat, power, water supply and wastewater disposal services, which use a standard break-even formula, actually encourages public utilities to overstate their costs, leading to an increase in the wear-out levels of municipal infrastructure assets.

One further reason for the high level of municipal infrastructure wear-out is the lack of access for public utilities to long-term investment resources. As a result they are not able to implement reconstruction projects for municipal infrastructure assets without significant tariff increases. Long-term investment and credit funds could allow public utilities to reduce the costs for their services and to provide credit recovery and a return on investments without significant tariff increases.

In order to increase the quality of public utility services and the effective use of natural resources, it will be necessary to undertake a large-scale implementation of reconstruction projects.

Reconstruction projects are in the strategic interests of the Russian Federation and will:

- provide more comfortable living conditions for the RF's population through improvements to the quality of their public utility services
- reduce the amount of dilapidated and emergency housing stock
- decrease power consumption by reducing losses during production and transportation
- provide more sustainable water use
- improve the state of the environment in the territories of the constituent parts of the RF

The total funding required for full recovery and reconstruction of the municipal infrastructure assets is 1.456 trillion rubles ($56 billion). This volume of investment cannot be met solely through the utilities' budgets - it is necessary to create the conditions to attract extra-budgetary funds. That is why this sub-programme is oriented towards making the municipal housing economy attractive to investors.

One of the main tasks for the sub-programme is the creation of the conditions to attract extra-budgetary funding sources for the reconstruction of municipal infrastructure assets.

Implementing the sub-programme will ensure:

- federal budget funds, budgets from the RF's constituent parts and local budgets are attracted into the reconstruction of municipal infrastructure assets
- the use of budget funds to implement reconstruction projects for municipal infrastructure assets
- the use of accessible funds from extra-budgetary sources for capital investment in municipal infrastructure assets
- stimulation of institutional reforms aimed at reducing the risks for extra-budgetary investment in reconstruction projects for municipal infrastructure assets, which will increase the accessibility for local authorities to extra-budgetary funds for the reconstruction of municipal infrastructure assets
- development of schemes to attract sources of extra-budgetary funding to the municipal housing economy

Within the framework of the federal task-oriented housing programme for 2002 to 2010 and the sub-programme (beside budget financing for the envisaged measures), the conditions will be created to attract sources of extra-budgetary funding to reconstruct municipal infrastructure assets. Projects in the different parts of the Russian Federation will be implemented on condition that implementation of measures for municipal infrastructure reconstruction in the constituent parts of the Russian Federation – the participants in the sub-programme – will bring about an improvement in the state of the municipal infrastructure. They must also result in improvements to the quality of public services. Institutional reforms undertaken within the framework of the sub-programme must ensure that sources of extra-budgetary funding are attracted into projects for municipal infrastructure reconstruction and that tariff increases within the public utility services are controlled.

The sub-programme is based on the following key principles:

- co-financing of reconstruction projects for municipal infrastructure assets, attracting both budget funds and extra-budgetary sources
- development of different forms of state and private partnership in order to attract sources of extra-budgetary financing to reconstruction projects for municipal infrastructure assets, using budgetary funds to reduce the investment risks
- open (non-discriminatory) selection of reconstruction projects for municipal infrastructure assets

The following challenges must be solved to enable these objectives to be reached:

- reconstruction of municipal infrastructure assets. Budget funds for implementing the sub-programme will be used to implement projects to rebuild municipal infrastructure assets related to the reconstruction of existing assets (those with high levels of wear-out), and also the construction of new assets intended to replace assets with a high level of wear-out

• improvements in the efficiency of the management of municipal infrastructure assets. This does not involve direct task-oriented budget financing, but its implementation will be assured by the choice of conditions for selection of the management company. Fulfilling these conditions will allow participation of the constituent parts of the RF in to participate in the company selection procedure. These companies will receive funds from the federal budget to implement municipal infrastructure asset projects. One of the most important keys to solving this challenge is perfecting the municipal housing economy's tariff regulation system. Another important element is engaging organisations with different ownership patterns on a competitive basis for municipal infrastructure asset management and the creation of concession-type agreements between local authorities and public utilities
• attracting extra-budgetary resources for financing the reconstruction of municipal infrastructure assets, including the development of credit facilities for these projects, will also be critical. The ongoing support of Municipalities and public utilities is envisaged during the process of obtaining credit for the reconstruction of municipal infrastructure assets.

Sub-programme measures will be enacted between 2006 and 2010 in two stages, the first from 2006 to 2007 and the second from 2008 to 2010.

Within the time frame of the first stage of the sub-programme's implementation it is assumed that financing for reconstruction projects will be obtained from federal budget funds on condition of obtaining co-financing from other sources. The majority of the funds for reconstructing the municipal infrastructure assets will come from the budget. As a result of implementing these measures the following goals should be reached:
• an up to 55% reduction in the wear-out level for municipal infrastructure assets
• an increase in the share of extra-budgetary financing of up to 20% of the total volume of investment into the reconstruction of municipal infrastructure assets
• an increase in the participation of private companies, which will manage the municipal infrastructure assets on the basis of concession agreements and other types of contracts for up to 15% of the total number of public utility services.

Within the time frame of the second stage of implementation it is assumed that financing from federal budget funds will be undertaken only when 45% or more of the investments needed for the reconstruction projects are financed from extra-budgetary sources in each constituent part of the Russian Federation. As a result of these measures, the following goals should be attained:
• a reduction of up to 50% in the wear-out level of the municipal infrastructure assets
• an increase in the share of extra-budgetary resources of up to 50% of the total volume of the investments into municipal infrastructure assets
• an increase in the share undertaken by private companies, which will manage the municipal infrastructure assets on the basis of concession agreements and other types of contracts for up to 30% of the total amount of utility services

Each year, the sub-programme's state customer will select the RF entities that are to participate in implementing the sub-programme events. activities. In order to take part in the selection process, the constituent parts of the Russian Federation must comply with the requirements for developing market relations in the RF's municipal housing economy and also with requirements for municipal infrastructure asset reconstruction projects.

Federal budget funds for implementing all projects within the Russian Federation will be given on condition of co-financing of reconstruction projects from the budgets of the RF constituents, local budgets and extra-budgetary sources (as set out in the stages of the sub-programme implementation):
• in 2006, the share of federal budget funds will be up to 42%, and the share of the RF constituent parts' budgets and local budgets will not be less than 33%, with the share from extra-budgetary sources not less than 25%
• in 2007, the federal budget's share of the funds will be up to 37%, with the share from the RF constituents parts' budgets and local budgets not less than 28% and the share from extra-budgetary sources not less than 35%
• in 2008, the share from federal budget funds will be up to 31%, the share from the RF constituent parts' budgets and local budgets not less than 24%, and the share from extra-budgetary sources not less than 45%

• in 2009, the share from federal budget funds will be up to 25%, the share from the RF constituent parts' budgets and local budgets not less than 20%, and the share from extra-budgetary sources not less than 55%
• in 2010, the share from federal budget funds will be up to 20%, the share from the RF constituent parts' budgets and local budgets not less than 15%, and the share from extra-budgetary sources not less than 65%

Funds from the budgets of the Russian Federation constituent parts, local budgets and extra-budgetary sources will include borrowed funds raised by the Russian Federation constituent parts, local authorities and private investors to implement municipal infrastructure objects. At the same time the share of budget funds coming from the Russian Federation constituent parts and local budgets can be reduced due to the increase in the share taken by private investors.

The right to obtain financing from the federal budget will be granted to reconstruction projects for water, heat and power supply networks and assets with a number of aims. These include reducing the amount of accidents, reducing water losses during production and transportation, increasing service life, reducing maintenance costs and undertaking reconstruction projects for water disposal systems aimed at improving the quality of the wastewater.

Reconstruction projects for municipal infrastructure assets can include construction of particular assets relating to environmental issues, primarily, for example, municipal waste processing plants. Priority will be given to reconstruction of those municipal infrastructure assets with the highest wear-out rate.

The pattern of financing specified will allow Municipalities and/or public utilities to borrow funds for the reconstruction of municipal infrastructure assets on a short-term, return and payable basis from financial establishments and at capital markets through issuing bonded loans.

During the period when the suggestions of the RF constituent parts are being evaluated in order to determine who will receive support from the federal budget within the time frame of the sub-programme, the state customer will use an integral evaluation criterion obtained by summing up of the following indices:

• the state of the municipal infrastructure
• the amount of investment in the municipal infrastructure assets
• planned investment in municipal infrastructure assets

The index relating to the state of the municipal infrastructure is determined as a ratio between the number of accidents in the previous year in the public utilities participating in the sub-programme and the length of municipal network that they serve.

The index relating to investment in municipal infrastructure assets is determined as a ratio between the actual capital investment in the municipal housing economy of the RF constituent part in the previous year from all finance sources, including charges for depreciation, to the average annual balance sheet value of the municipal housing economy's fixed assets within the constituent part of the RF.

The index of planned investment in municipal infrastructure assets is determined as a ratio between the cost of reconstruction projects of municipal infrastructure assets projects specified in the application from the RF constituent part and the average annual balance sheet value of the fixed assets of the public utilities participating in the project.

This integral evaluation criterion reflects the seriousness of the problem of reconstructing the municipal infrastructure assets and the possibility of solving it by raising budgetary and extra-budgetary funds.

The RF constituent parts will be the participants in the sub-programme for the current year, as they take the first places under the evaluation (integral) criterion. The number of RF constituent parts that can participate in the sub-programme will be determined on the basis of allocated budget funds, and also the amount of financial support that each can provide that yields the greatest value under the integral evaluation criterion. A list of sites and assets for the selected construction projects within the time frame of the sub-programme will be formed according to established procedures and approved by the government of the Russian Federation.

Each quarter, the RF constituent parts and local authorities, whose territorial asset reconstruction projects are being implemented using federal budget funds within the sub-programme, will submit reports about the use of federal budget funds and implementation of the projects according to the procedure established by the state customer.

Selection of reconstruction projects for municipal infrastructure assets and also of the Municipalities to participate in the implementation of the sub-programme by the RF constituent parts will be performed in accordance with the order established by the legislation.

The executive authorities of the RF constituent parts, together with local authorities, will prepare reconstruction projects for municipal infrastructure assets in order to submit them for selection within the framework of the sub-programme.

Successfully achieving the sub-programme will allow the following by 2010:

- a decrease in the wear-out rate for municipal infrastructure assets from 60% in 2004 to 50%
- growth in extra-budgetary funding sources in the reconstruction of municipal infrastructure from 12% in 2004 to 50%
- an improvement in the quality and reliability of public utility services
- an improvement in the state of the environment in the RF constituent parts and the Municipalities
- creation of a sustainable institutional basis for private sector participation in financing reconstruction projects for municipal infrastructure assets and municipal infrastructure management

4. Private sector participation in management of the water supply and wastewater disposal systems

4.1 General information

In 2000, the participants in the Consultation summit held by the Ministers of Economic Affairs / Finance and Environment Protection of the New Independent States (NIS), formed after the disintegration of the USSR, held in Alma-Ata, adopted the Guidelines for reforming the water supply and sewerage systems of NIS. These specify that 'private sector participation can contribute to improving the efficiency of fund raising, and the introduction of new technologies to the activities of the Vodokanals'.

In 2003 and 2004 a number of private operators working in the public utility services market for water supply and wastewater disposal were founded in Russia. These are:

- Rossijskiye kommunalnyje sistemy, OJSC (RKS)
- Novogor-Prikamye, CJSC
- Evrazijskoye vodnoye partnerstvo, LLC (EVP)
- Rosvodokanal, OJSC (RVK)

In 2005 Don-Yug and Don-Sever, companies that have foreign capital involvement (ABB, of Switzerland), joined them. They implement reconstruction projects for the Vodokanals in the Rostov region and the Stavropol territory.

In addition, several Vodokanals (for instance in Syzran, in the Ulyanovsk region and in Krasnogosk in the Moscow region among other cities) were corporatised with private sector participation.

Private operators provided water supply and sewerage services in cities with populations of:

- two million people (less than 2% of the country's population) in 2004
- nine million people (less than 6% of the country's population) in 2005
- 15 million people (more than 10% of the country's population) in 2006 (estimated)

It is expected by 2010 that private operators will serve water supply and sewerage systems in cities for a population of not less than 40 million people, 30% of the country's population.

The development of this business at present mainly involves private operators winning contracts for water supply and sewerage systems in cities with populations of not less than 200,000 – the administrative centres of the RF constituent parts. It is expected that later on these companies will gain control of other towns in regional centres, where they are on a secure footing. That means that development of private sector participation in the water supply and sewerage sector in Russia will most probably be organised along territorial lines (each will work within a set territory).

In 2004 the Organization of Economic Cooperation and Development (OECD) prepared a review of private companies working in the Russian water supply and sewerage services market. The OECD undertook a questionnaire survey of the Vodokanals; a summary of the results is given below.

The basic reason for their entry into the public utility business, the companies explained, was the volume of the market. They also cited the availability of guaranteed sales and the considerable potential for efficiency improvements and cost reductions.

All of the companies declared their desire to do business on the basis of long-term lease contracts for municipal infrastructure assets. They also undertook to provide sufficient investment obligations - from $50 to $500 million within each five year period (authors' note: from $15 to $20 within each five year period per head of population served). At the same time, all of the companies are ready to take part in tenders to finalise such contracts.

Many of the companies included in the review stated that 'with skilful management even the current tariffs allow sufficient running to cover maintenance and even a slight improvement in the current state of the fixed assets transferred into its management'. Rosvodokanal and Novogor-Prikamye assert that even the imperfect legislative and regulatory framework and budget financing shortfalls have not had a significant influence on their activities.

The paradox within these statements is that the inadequate tariff policy and budget financing for the so-called 'falling out income' are considered to be the main factors in the unstable development of Russia's municipal economy over the past 10 years. The key to this paradox is partly contained in the fact that all these companies act on commercial principles and as a result take the set tariffs as rigid budget limitations.

Nevertheless, as can be seen from global experience, private sector engagement requires certain (quite high) requirements from municipal authorities. The more functions that are given to the private sector, the more complicated are the procedures for selecting a private operator, monitoring the fulfilment of their functions, and regulating their tariffs. And, above all, private sector involvement needs an understanding on the part of local authorities of the business and set-up principles for sustainable partnerships between authorities and business in municipal housing.

Unfortunately, the majority of Municipality administrations in Russia have had a shortage of qualified personnel for quite a long time. Participants in the review specified this as a problem. As a result private sector engagement in the supply of public services is a political decision in most cases, in which importance is placed on random and subjective factors. At the same time contractual relationships between municipal governments and private operators mainly appear to be agreements of intent rather than full contracts to delegate management. It is necessary to note that private initiative involvement on the basis of management contracts that do not impose major responsibilities on the private sector still provides a remarkable improvement within a short period (one to two years). The arrival of 'private owners' leads to an improvement in performance (a reduction in the incident rate and water losses) and the financial state of the Vodokanals.

It is important to note that private companies create a drive, which is financially motivated, to change the existing institutions of engineering infrastructure management. The question is, above all, about the conditions of contractual relationships between the public utilities and their services and the municipal governments (the owners of the municipal infrastructure), and about the division of responsibility and risks in such contracts. The question is also one of the interest of large utilities in the legislation and their contribution towards unified, honest 'rules of the game'.

In this way, formal and more transparent relationships are created in the sector, political interference is reduced and therefore additional motivations are created for achieving efficiency and sustainable development within the sector.

4.2 Obstacles in the way of private sector participation

4.2.1 General description of the problems

The OECD review also contains a list of obstacles that the Russian private companies faced when implementing their projects in the various Municipalities. Many of these obstacles are connected directly with the entry of private organisations into the sector and reflect deep-seated and unresolved problems within it, which require immediate resolution. We would like to stress the problems, which are mostly not peculiar to the CIS countries. They are:

- considerable political risk, a desire to manipulate the public utilities and a historical unwillingness to approve economically-sound tariffs
- non-transparent tariff regulation
- the absence of long-term credits
- the crude nature of the legal basis for long-term investment
- personnel problems
- problems with the assessment and registration of property transferred to new management

Below is a brief description of each of the problems listed above.

4.2.2 Political risks related to the cancellation of contractual relationships and non-fulfilment of

obligations by authorities
As mentioned earlier, private sector involvement in MHE management in Russia is in most cases a political decision. This seems to be true not only for Russia but in practically all cases where private businesses enter into this sector. In general it is considered that 'a fixed political commitment in favour of project implementation is a required condition for the formation of a state-private partnership'.

One of the most serious risks, which the private sector in Russia faces, is the reliability of the answer to the question of whether local authorities are going to participate in the sector's work and how they are going to do this. Are the authorities going to fulfil their obligations? Will they change their goals? Will they cancel the contract?

Unfortunately, in Russia's local administrations short planning and decision-taking horizons (one to two years) still prevail. That is why it is difficult to be sure that not only when authorities change, but even with the old authorities that signed the contract with the private company, they will have the 'fixed political commitment' to fulfil their obligations.

Political risks relate not only to tariff policy. Municipal government, as the owner of municipal infrastructure, can fail to fulfil its obligations in terms of capital investment, for financing reconstruction works (this is particularly important in lease contracts) or in terms of covering the company for the so-called 'falling out income' which relates to customer privileges relating to payments for public services.

The considerable political risks were the factor that determined the unwillingness of many foreign private operators to enter the Russian public utility services market.

4.2.3 The availability of unregistered property
Years of economic mismanagement have led to a situation where municipal governments do not know the true state of their fixed assets and do not even have documents formally confirming their ownership rights to some of the assets. At the same time, the transfer of municipal property into long-term lease agreements requires its state registration. This can take from one to three years depending on the size of the city; it is a serious (though temporary) obstacle to the conclusion of long-term contracts between municipal governments and private businesses. As a result most of the Russian private companies had to opt for short-term contracts for the lease of municipal property, as this does not require state registration of the contract and leased property.

4.2.4 Inadequate tariff regulation
Tariff regulation provides a regulated public utility with the revenue it needs for the sustainable financing of its activities, and does not allow it to abuse its monopolistic position. Nevertheless, the situation in which practically all of the population and most consumers – the potential electorate – use the centralised water supply and wastewater disposal systems, often leads to politicisation of the tariff formation process.

The Russian experience is absolutely identical to the international one in the sense that the political dependency of a regulatory body quite often leads to a situation where, with populist aims, it may refuse a water supply company's (regardless of whether it is a private, state or municipal company) request to raise tariffs, even if the request is justified by price rises and consequent growth in the company's costs.

It is necessary to acknowledge that the basis for the politicisation of the tariff formation process in Russia is the low ability of the population to pay, which limits tariff growth. Thus, attempts to return investments solely through tariffs can be impractical. The problem is aggravated by the lack of long-term loans in the Russian capital markets.

4.2.5 Lack of long-term financing
The transfer to a private operator of the responsibility for investing in the capital-intensive fixed assets of municipal infrastructure creates as a key factor the possibility of finding creditors or investors willing and ready to offer the required funds. In Russia, private companies have faced a major dearth of long-term loans. Russian banks do not give long-term credits, and it is not possible to get money from pension funds that have idle investments.

One possible solution to the problem is to involve potential investors in the negotiation process between the authorities and the business at the earliest stages of a project, during the development of approaches to reforms in municipal infrastructure management.

Global experience proves that the participation of the interested parties at the earliest stages of a project is key, though it introduces additional costs and complicates the process as a whole, but in the later stages it brings essential positive results.

4.2.6 Weaknesses and imbalances in local budgets

New approaches to local government organisation are stipulated in the Federal Law of 6 October 2006 No 131- FZ entitled 'About general principles of local government organisation in the Russian Federation', which comes into force from 1 January 2009. At this time, the income of local budgets will be sufficient to enable full implementation of the local authorities' functions and tasks, including organisation of water supply and wastewater disposal within their territory.

At present, the income from Municipality budgets is not sufficient to cover all of the expenses of the municipal government. That is why non-fulfilment of local authority obligations may not be connected with so-called 'opportunistic behaviour' towards private investors, but rather with real weakness and imbalances within local budgets, and difficulties in making long-term forecasts because of this unstable income.

In terms of the budget financing shortfalls, the question is not only about the failure to pay the 'falling out' income to companies, which relates to payment of housing and municipal services (HMS) and subsidies, but also about the failure to pay against guarantees issued. As an example, these can be set during the conclusion of a loan agreement between a private operator and a bank and paid in the event of non-fulfilment or incomplete fulfilment of obligations by a loan debtor.

The weakness of local budgets also includes a risk of default by the local authorities to a private operator in regard to capital investments into reconstruction and development of infrastructure. This is particularly the case when an operator concludes an agreement for the lease of municipal infrastructure assets, when the lease payment (which previously was given to the Vodokanal and was used by it for replacing its fixed assets) becomes part of the income of the Municipality. If there is a budget deficit (that is, if the expenditure budget exceeds the income budget) financing for the replacement of fixed assets may not be secured, meaning that an operator may not be able to fulfil its contractual obligations to maintain these assets in a good state.

4.2.7 Underestimating the risk of investment cost demands

During the conclusion of a lease agreement an operator certainly does not know all of the information about the true state of the leased fixed assets, as the existing reporting system in Russia does not make it possible to obtain realistic information for evaluation before the municipal infrastructure is taken into its management. The absence of realistic information can involve companies in the significant expense of an audit, which can reveal much larger investment needs than were anticipated during the conclusion of the lease agreement.

If, under its contract with the municipal government, a private operator takes on responsibility for the reliability, safety and performance of the water supply system, then it, naturally, will try to estimate the cost of fulfilling these obligations. But what can it do if the state of the system turns out to be worse then was supposed by the operator, and its additional expenses are not covered by income? What can it do if the municipal government refuses to review its tariff? In this case either the project's profitability will be lower than planned or the operator will not fulfil its obligations, providing an opportunity for early termination of its contract, which can lead to an increase in its losses.

If the municipal government does agree to include respective additional expenses within the tariff, the new tariff can be unacceptably high and payment collection can decrease. In addition, the population and authorities will suspect that the operator is trying to cheat them, to increase its income and abuse its monopoly powers.

This uncertainty has to be reduced before concluding a lease or concession contract. For this purpose local authorities must organise an independent audit and evaluate the investment requirements for renewal, upgrading and reconstructing their fixed assets.

4.2.8 Low personal income and affordability of tariffs

One of the most important issues for private companies around the world is how quality improvement and service availability should influence tariffs. To be more exact whether these improvements can be financed solely through the savings made from efficiency improvements (that is, cost savings at a given level of tariffs), and if it not, whether the consumers can afford this and whether they will agree to pay more.

In other words it is likely that a tariff that allows all justified expenses for operation and maintenance of the fixed assets to be covered may be unacceptably high because of the low income of the population and budget weaknesses, which is particularly important in small Municipalities.

The problem is that taking into consideration all the above factors (the imbalance and weakness of local budgets, the lack of information about the current state of fixed assets, and so on) the long-term financial sustainability of private operators can be called into question, as the investment needs of the sector can be significant. Bear in mind, there have been no large capital investments or adequate maintenance and repair over the last 10 to 15 years).

4.2.9 Conclusion

Any private operator expects to conclude a lease or concession agreement. At the same time, the investment obligations of the lessee cannot be confirmed by an objective estimate of the investment risks and construction of realistic finance model, which complicates the ability to obtain long-term credits. The solution to this problem can be found in regulation, envisaging the authorities having a function to perform independent technical, process and financial estimates before transferring infrastructure assets to leases or concessions and to provide all interested parties with the objective information they need.

We would like to note one positive peculiarity in the incipient infrastructure asset management market in Russia. Many of the companies participating in this market have opted for a policy of openness in information provision. This leads to transparency in relationships and better understanding between key process participants of management procedures, providing a more objective development of strategic plans, monitoring of goals and achievement of tasks.

4.3 Rossijskiye kommunalnyje sistemy, OJSC (RKS)

Rossijskiye kommunalnyje sistemy, OJSC (RKS) is one of the first projects in the country set up to reconstruct the management systems for municipal infrastructure belonging to the Municipalities. RKS OJSC was formed in May 2003. The mission of the company was the formulation of a standard level of quality for public utility services. Its strategic and priority task was the raising of large-scale investment in the municipal economy of the Russian regions.

RKS is positioned as a strategic investor oriented towards the development of management and technological competencies, and the introduction of investment planning practices to absorb interest created by the reduction in expenses in the sector. The priority task of the RKS leaders is to reduce administrative and technological costs within the enterprises under its control.

The company understands that it is doing business in a sector where technical problems have remained unsolved for decades because of a shortage of funds. RKS experts calculated that when municipal infrastructure assets are taken into management, the first two to three years should be spent on legal procedures (a technical inventory and state registration of the property) and restoring the reliability of the network and facilities. Only after that can serious business be undertaken.

At present the main type of work that RKS is involved in consists of short-term lease contracts (of 11 months) in cities with populations from 25,000 up to one million.

The first long-term investment agreement (for 15 years) was signed by RKS on 28 May, 2004 in Kirov. The estimated amount of investment that needs to be put into the city's water supply system is $20 million.

The main type of RKS participation in the management of cities' housing economies is a combination of a long-term infrastructure lease with a municipal order for its repair and improvement. Issues of financial recovery and raising investment can be solved only on the basis of contracts that provide the operator with a long-term flow of funding.

One new RKS business model involves an evolution in its lease relationships developed during the first year of working under the conditions of its 11-month lease contracts and negotiations with municipal governments. It is the closest to the definition of a concession contract - a combination of a labour contract through municipal order that is transferred into a lease contract when improvements are made by an operator from extra-budgetary sources that are paid in long-term instalments by the municipal government from receipts from actual lease payments. The local authorities formulate the requirements in light of the municipal order. All assets and associated improvements undertaken during execution of the contract are transferred into the ownership of the local government.

The stages of the main plan for RKS's work is given below:

- in the region's administrative centre (the RF constituent part) an RKS subsidiary is founded, for

example, Tambovskiye kommunalnyje sistemy, OJSC
• the Municipal Unitary Enterprise – Vodokanal - transfers its property into the treasury of the Municipality, after this its liquidation begins
• the RKS subsidiary takes over municipal infrastructure assets in a short-term lease and after property registration it either undertakes water supply and wastewater disposal in its own right or creates a new legal entity - a Vodokanal

RKS cooperates closely with municipal governments during the inventory, technical and economic audit and state registration of the municipal infrastructure, because without this registration it would be impossible to conclude any long-term lease or other investment agreements.

The water supply and wastewater disposal systems (Vodokanals) managed by RKS (as of January 2006) are:
• Volgograd (Volgograd region) – 1 million population
• Tomsk (Tomsk region) – 487,000 people
• Kirov 450,000 people
• Tambov (Tambov region) – 287,000 people
• Petrozavodsk (Karelia) – 266,000 people
• Blagoveshchensk (Amursk region) – 218,000 people
• Kachkanar (Sverdlovskaya region) – 44,000 people

4.4 Evrazijskoye vodnoye partnerstvo, LLC (EVP)

4.4.1 General information

Business:

Implementing state-private partnership projects, undertaking property development, operating and managing company functions

Business objectives:

Vodokanals – water supply and water disposal companies of medium and large cities

Business territory:

The Russian Federation and Ukraine

Activities:
• taking the role of the Vodokanal operator
• business restructuring and efficiency improvement
• management, development and repair of water supply and wastewater disposal infrastructure
• development of investment projects

Target:

Gaining a business advantage by improving the level of satisfaction felt by consumers through provision of an uninterrupted, high quality service that is affordable and available

Short-term objectives (up to 2008):
• the conclusion of long-term contracts in seven to 10 cities with a total population of 5 million and a turnover of $200 million per year
• the formation of an investment project portfolio worth $100 million

Tasks:
• optimising the Vodokanals' business activities, including management, financial, commercial, economic and technical aspects
• managing the Vodokanals' systems and facilities in order to enable their replacement, recovery and maintenance
• raising long-term investment
• developing joint projects between the state and private sectors for the effective management of municipal infrastructure

Authorised capital of the company:

10 million rubles ($0.4 million).

4.4.2 Strategy

The commercial objectives of the company will be reached through the following sequence of actions:

- fixing and increasing the company's value by concluding long-term business contracts and improving the Vodokanal's commercial efficiency
- providing the target market with data about the value of the company (its profitability, project portfolio, strategic partners, and the participation of international financial institutions and other large investors
- sale of the company's shares on the western financial market

The cost of the company is determined as the sum of the following components:

- NPV (net present value) of projects managed or controlled by the company
- estimated value of the company's knowledge relating to organisation and business development
- the value of the company's assets
- an estimate of other factors

4.4.3 Main activities in 2004 and 2005

To realise its strategic objectives the company undertook the following activities during 2004 and 2005:

- the company was established, and qualified personnel with considerable competence in management were recruited
- individual schemes for interacting with Municipalities, Vodokanals, potential investors and other stakeholders were created and implemented
- in order to obtain credit on the Russian and international financial markets a mechanism to provide guarantees of security and surrender was developed in detail and its practice began (in particular, at present a mechanism for receiving credit from the Vneshtorgbank and the EBRD is in place).
- EVP LLC took over the management of OmskVodokanal, OJSC (the population of Omsk is 1.14 million) and the Vodokanal, OJSC of Rostov-on-Don (population 1.1 million)
- short-term and long-term investment programmes were developed for reconstructing and upgrading the water supply and sewerage assets of Omsk and Rostov-on-Don
- a technological audit of OmskVodokanal, OJSC was undertaken by seven independent international consulting companies
- a tender for developing an investment project to upgrade and reconstruct Omsk's water supply and sewerage systems worth €30 million ($41 million) was announced and organised
- a training system for the personnel of the company and the Vodokanals it controls was set up
- formation of corporate processes, methods and forms of business management, business operation and development took place within the company

4.4.4 Project OmskVodokanal, OJSC

EVP undertakes sole executive body functions in terms of the decisions of the General Meeting of Shareholders and the management contract concluded by EVP with OmskVodokanal, OJSC.

Issues of interest at OmskVodokanal, OJSC:

- there is a high level of technical innovation: vibration diagnostics for pumping units and motors, use of a SCADA system, modern pumps, use of PVC pipes and use of trenchless technologies for pipe replacement
- there is an adequate number of employees (2,000)
- the Vodokanal had developed an outsourcing initiative for the servicing for its wastewater pumping stations and the maintenance of its water supply and sewerage assets

Existing problems:

- there is not sufficiently effective management of consumer demand and service provision
- there is an absence of metering devices in apartment buildings
- there is ineffective management of the distribution network
- there is a high level of wastewater collector system wear-out

Current state of affairs:

EVP was awarded the contract for the long-term development of the water supply and wastewater disposal infrastructure. With implementation of the contract (a long-term lease agreement for 25 years) an investment agreement for funding for water supply and sewerage assets valued at 450 million rubles ($17.5 million) over three years (to April 1, 2008) was concluded alongside development of a long-term investment programme worth around four billion rubles ($155 million). At present more than 100 million rubles ($3.9 million) has been transferred for upgrading and developing the municipal water supply and sewerage systems.

4.4.5 Project Vodokanal, OJSC in Rostov-on-Don

EVP undertakes the functions of the sole executive body in relation to decisions of the General Meeting of Shareholders and the management contract concluded by EVP with Vodokanal, OJSC in the city of Rostov-on-Don.

Issues of interest:

- the Vodokanal, OJSC in Rostov-on-Don (Vodokanal) serves over one million residents, and has an annual turnover of $30 million
- in 2005 the Vodokanal received an IBRD credit for the development and repair of the municipal water supply system worth $10 million
- in 2005 the Vodokanal received a grant from the Global Environment Facility (GEF) for constructing a treatment system for the phosphorus and nitrogen components of the wastewater discharged into the river worth $10 million
- the Vodokanal has a high level of technical and management expertise

Existing problems:

- the unsatisfactory state of its facilities and networks
- the absence of adequate financing
- a high level of unaccounted-for water

Current state of affairs:

In September 2005, the Municipal Unitary Enterprise (MUE) Vodokanal of Rostov-on-Don was corporatised as the Vodokanal of Rostov-on-Don, OJSC, 100% owned by the Rostov-on-Don city administration. Engineering infrastructure assets (networks, water intake and water treatment facilities and pumping stations) were not included in the authorised capital of the joint stock company and were transferred into the lease of the Vodokanal, OJSC in Rostov-on-Don.

In November 2005, by decision of the General Meeting of Shareholders, the functions of the sole executive body were entrusted to EVP, which concluded a management contract with the Vodokanal, OJSC in Rostov-on-Don.

In 2005 EVP LLC invested over 300 million rubles from its own funds in developing the water supply for the city of Rostov-on-Don, which led to the President of EVP, Sergey Yashechkin, and General Director of the Vodokanal OJSC, Boris Persidsky, being awarded citations from the RF President.

4.5 Novogor-Prikamye, LLC (new city infrastructure of Prikamye)

4.5.1 General information

Novogor-Prikamye, LLC is a subsidiary of Novogor, CJSC, established to implement investment projects in the housing economy sector in the Perm region. Novogor, CJSC is a special-purpose subsidiary of Interros CJSC set up to implement projects in the housing sector. It was created on 1 December 2003.

Note: Interros is one of the largest private investment companies in Russia. The market value of the assets under its control exceeds $10 billion. At present, Group enterprises work in all of the regions of Russia and also in a number of countries in Europe, Asia and North America They produce about 1.3% GDP and employ over 190,000 people.

Novogor-Prikamye's mission is to improve the living standards of Perm's population by providing high-quality services for water supply and wastewater disposal at a reasonable price.

The company's strategic objective is to raise the water supply service level in Perm to that of European countries within 12 to 15 years.

Business development strategy: to be a client-oriented organisation that re-engineers the service from manufacturers to the company undertaking the services. The basis of Novogor-Prikamye's strategy is the administrative reform business model - a division of the company's departments into regulatory, management and executive roles.

Novogor-Prikamye has developed a forward strategy for Perm's water supply and wastewater disposal systems to 2020; its key principle is aligning itself with the performance of utilities in European cities that have similar infrastructures and population sizes. In order to prepare its development strategy and technical audit, the company engaged leading specialists Severn Trent Water International (of the UK).

During Novogor-Prikamye's work in its management company capacity:

- modern management, management and business accounting approaches are being introduced and used
- the Permvodokanal municipal enterprise's accounts receivable and payable are being reduced
- the percentage of payments collected is being increased
- the company's expenses are being significantly reduced
- the scope of its refurbishment work has been reviewed and expanded by 1.5 times

The General Director of Novogor-Prikamye, LLC is one of 12 directors of leading enterprises within the region that are included in the Economic Council of the Governor of the Perm region.

4.5.2 Key performance indicators for 2004

See table 4.5.2.

4.5.3 Main financial and economic indicators for 2004

See table 4.5.3.

Table 4.5.2. Key performance indicators for 2004.

	Indicators	Units	2003	2004 plan	actual value	% v 2003	% v plan
1.	Water lift	thous.m³	169,978	169,922	167,150	98.3	98.4
2.	Purchase water	thous.m³	9,336	8,796	8,797	94.2	100.0
3.	Consumption for own process needs	thous.m³	8,052	9,257	8,467	105.2	91.5
4.	Supply to mains	thous.m³	170,740	169,092	167,067	97.8	98.8
5.	Pumped water	thous.m³	190,225	175,858	168,604	88.6	95.9
6.	Useful output (7+8+9)	thous.m³	132,862	132,348	130,048	97.9	98.3
7.	including sale to:	thous.m³	109,077	108,502	106,384	97.5	98.0
7.1.	Population	thous.m³	82,672	82,286	81,561	98.7	99.1
7.2.	Budget (regional + federal)	thous.m³	4,447	4,825	3,781	85.0	78.4
7.3.	Municipal budget	thous.m³	5,133	4,974	4,884	95.1	98.2
7.4.	Other companies	thous.m³	16,825	16,417	16,159	96.0	98.4
8.	Own needs (domestic)	thous.m³	1,323	1,110	1,023	77.3	92.2
9.	Water losses (for sale)	thous.m³	22,463	22,736	22,641	100.8	99.6
10.	Water leaks (4+6)	thous.m³	37,878	36,745	37,018	97.7	100.7
11.	% of leak from supplied water	%	22.2	21.7	22.2	99.9	102.0
12.	Waste water disposal (13+14+15)	thous.m³	137,982	136,145	135,374	98.1	99.4
13.	including sale to:	thous.m³	115,564	113,536	112,558	97.4	99.1
13.1.	Population	thous.m³	82,220	81,759	81,634	99.3	99.8
13.2.	Budget (regional + federal)	thous.m³	4,200	4,058	3,625	86.3	89.3
13.3.	Municipal budget	thous.m³	5,192	4,933	5,044	97.2	102.3
13.4.	Other companies	thous.m³	23,954	22,786	22,255	92.9	97.7
14.	Own needs (domestic)	thous.m³	229	203	230	100.4	113.3
15.	Unaccounted waste water consumption from population	thous.m³	22,189	22,406	22,586	101.8	100.8
16.	Biological treatment of waste water	thous.m³	93,217	96,281	95,705	102.7	99.4
17.	Daily average supply	thous.m³	468	462	456	97.6	98.8
18.	Daily average sale	thous.m³	521	480	461	88.4	95.9
19.	Daily average losses	thous.m³	104	100	101	97.5	100.7
20.	Daily average waste water disposal	thous.m³	378	372	370	97.8	99.4

4.5.4 Development strategy for the water supply system

The goal of the development strategy is to improve the living standards of the population in Perm

Indicators		2003	2004	Change
Revenue, total	mln. rub	934	996	+ 62
Including – water supply	mln. rub	491	521	+ 30
- water disposal	mln. rub	443	475	+ 32
Tax amount (no tax arrears)	mln. rub		292.48	
Staff on the payroll	persons	1,790	1,741	- 49
Average wages (no arrears)	rub/person	5,782	6,345	+ 9.7%

Table 4.5.3. Main financial and economic indicators for 2004.

by raising the performance of its water and sewerage system to that of developed industrial cities in western Europe with matching water supply infrastructure and population sizes.

The basis of the development strategy includes an assessment of the current state of the fixed assets and service levels based on a system of key performance indicators. In all, 26 key performance indicators of production activities have been developed.

This assessment can be represented by means of a 'spiderweb' diagram (see Figure 4.5.4), which shows the values of the key performance indicators in seven key groups:

- availability and quality of drinking water (this indicator is highly important because of the need to comply with set water quality standards and also the frequency of water sampling).
- an uninterrupted water supply (this indicator is also highly important, which is evident because in providing a 24-hour service for water supply and wastewater disposal, a 20% deviation is currently allowed, in particular for water supply failures in remote districts).
- consumer service (work with consumers) – the importance of this indicator is explained by the lack of initial data and also by the differences in the basis for comparison (one example is the availability of metering devices related to the number of connections)
- environmental safety – this indicator relates to the availability of emergency discharge facilities for untreated wastewater.
- financial stability – this indicator is influenced, above all, by the existing low energy efficiency and high incident rate.
- risk of terrorist acts - this indicator is determined by the system's ability to maintain stable production in the event of the power supply being cut off and also by the time taken to recover the water supply system (reservoir capacity).
- training and advance vocational training

This comparison gives clear picture of main development trends of Vodokanal and it is the basis for strategy formation of water supply and water disposal system. Also specified indicators are control instruments for annual operating results of Novogor-Prikamye by the city.

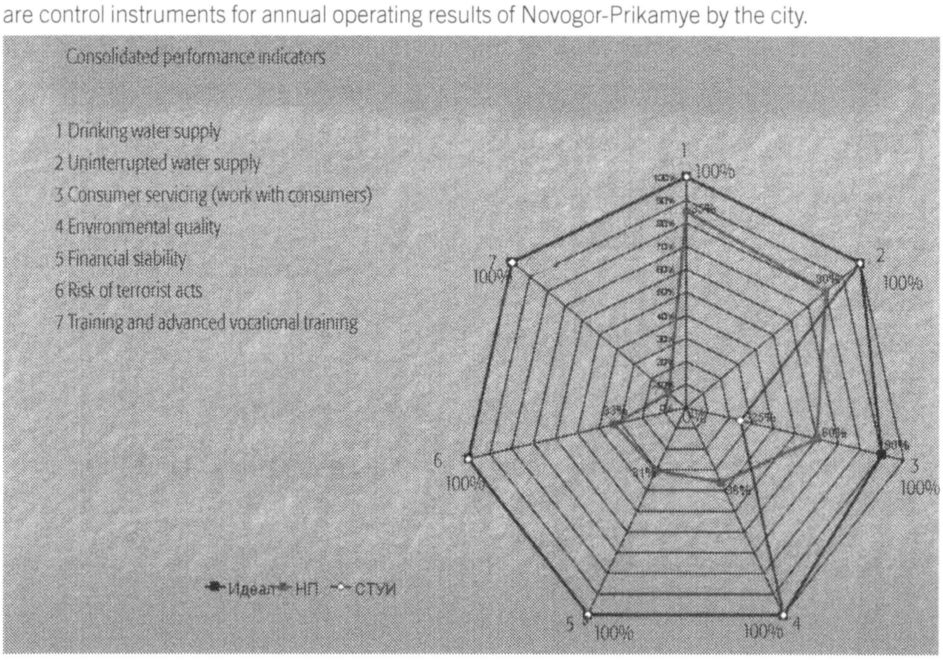

Figure 4.5.4. Consolidated performance indicators for Novogor-Prikamye, LLC.

4.5.5 Investment programme

The formulation and implementation of Novogor-Prikamye's development strategy for is water supply and wastewater disposal systems is based on the following key components:

- an asset management plan
- new technologies
- a technology strategy
- programme management

The asset management plan is the basis for capital investments, as most of the investment programme will be devoted to the issues surrounding replacement and renewal of existing production facilities.

New technologies

One of the aspects of implementing the investment programme is the introduction of new technologies during the reconstruction and renewal of production facilities. In particular, it is assumed energy-efficient technologies will be introduced – where required, variable speed drives and devices to enable a reduced-current start-up. In order to improve process efficiency (excluding human factors) it is assumed that a local Automatic Process Control System and remote control system (SCADA) will be introduced.

The total amount of financing for projects within the investment programme is $315 million over 16 years (on average around $20 million per year).

Main sections of the investment programme:

- water abstraction and treatment – $63.8 million
- water transportation and storage – $63.1 million
- water distribution – $31.2 million
- wastewater disposal – $48.3 million
- wastewater treatment – $104.4 million
- other projects (renewal of the laboratory and the communication system - creating a LAN and so on) – $4.3 million
- total: $315.3 million

The main focus areas of the short-term investment programme for 2005 to 2006 are the following:

For the water supply:

- improvements to the water supply to remote districts of the city, including reconstruction of the Southern pumping station (increasing reservoir capacity, reconstructing pumping units)
- creating a hydraulic model of the water supply and wastewater disposal system operation to provide a justification based on the calculations of the technological decisions taken, and also to introduce a production record system.
- re-laying sections of the network that are in a critical state at a rate of 2% per year of the total length.

For wastewater disposal:

- renewal of trunk sewers at a rate of 2% per year of the total length
- reconstruction and extension of the existing biological wastewater treatment facilities to a capacity of 420,000m^3/day.

Management of the investment programme

The investment programme submitted for approval to the city envisages joint management of the operation. In particular, there are plans to set up a technical council at Novogor-Prikamye, LLC involving the participation of the city administration, to work out the technical policy and organisation for the coordination committee on capital investments within the city administration, in which representatives of Novogor-Prikamye, LLC will participate.

4.5.6 Development programme for Novogor-Prikamye, LLC from 2005 to 2009

a. Target

In accordance with the development strategy for Perm's water supply and water disposal system to 2020 (the development strategy), administering the utility is the present short-term (five year) planned task.

The main objective of the present planned task is to focus the work of Novogor-Prikamye, LLC on the following main areas:

- stabilising the water supply in remote districts of the city and maintaining the water supply in central districts at the stipulated level
- providing environmental safety for the people of Perm by reducing the pollution in the Kama river caused by untreated wastewater, improving the quality of wastewater treatment, and achieving set standard values for wastewater discharges to the Kama

b. Novogor-Prikamye, LLC's planned task

Sampling point	Chemical analysis (% of non-standard samples of total quality)	Bacteriological analysis (% of non-standard samples of total quality)
Dzerzhinsky district	8.3	3.6
Industrial district	3.9	7.1
Kirvsky district	12.4	7.6
Leninsky district	7.5	2.5
Motovilikhinsky district	4.9	9.4
Ordzhonikidzevsky district	0.5	3.1
Sverdlovsky district	5.3	15.5

Table 4.5.6. Fixed drinking water indices.

In line with the planned task, Novogor-Prikamye, LLC will develop a production programme each year.

The annual production programme is approved by Perm's city administration. Responsibility for fulfilling the production programme is taken by Novogor-Prikamye, LLC. Monitoring the fulfilment of the production programme is undertaken by the MHE department of Perm's city administration.

In the event of real issues arising, the Perm city administration or Novogor-Prikamye, LLC are entitled to suggest introducing modifications to the planned task, with a justification of the modification of the conditions or volumes set under the present planned task. All modifications to the current planned task must be introduced only with the consent of all parties involved.

A modification of the planned task will be the basis for Novogor-Prikamye, LLC to undertake a production program review. Planned task modifications are confirmed by intermediate acceptance certificates relating to each modified point.

c. Efficiency measures

In accordance with the development strategy measures within the planned task, efficiency is focused on five key areas:

- availability and quality of drinking water
- an uninterrupted water supply
- consumer service
- environmental safety
- effective use of leased property

These indicators are evaluated by monitoring of Novogor-Prikamye, LLC, undertaken in accordance with the general agreement. The following target values are envisaged.

Availability and quality of drinking water

The plan foresees that there will be no deterioration in quality in terms of chemical and bacteriological indices within the city's distribution systems. The fixed indices are shown in Table 4.5.6.

Uninterrupted water supply – the plan foresees:

- development of design and budget documentation in 2006 for the city's water supply in full compliance with the standards; stage-by-stage pressure stabilisation in district water supply networks, ensuring a guaranteed minimum free head at the intakes for all 4,900 assets connected to the water supply system by 1 December 2009:

- for single-floor dwellings (including CHPS) not less than 10m head
- for multiple-floor dwellings a 4m head for each next floor (up to the fifth floor in the case of direct connections)
- provision of storage capacity for drinking water in the city by 1 December 2009 in accordance with SNiP 2.04.02-84

vregistration of all of Perm's water supply and wastewater disposal networks (excluding the networks of industrial consumers) into a long-term lease by 1 December 2009

- re-laying and disinfection of the water supply and wastewater disposal networks at a rate of 44km per year
- full certification for the water supply network facilities and structures by 2007
- creation of a full-scale hydraulic model of the city's water supply and wastewater disposal system by 2009
- construction of a dispatch system for the city's water supply and wastewater disposal assets (a main transfer pumping station) by 2008;
- development of a technical policy and engineering supervision procedures for the construction of engineering infrastructure for the city's water supply and wastewater disposal system and their submission for approval by the head of the city within six months of the planned task receiving approval
- ending the systematic excess period allowed for dealing with emergencies, in agreement with the MHE Department of Perm;
- development of design and budget documentation for provision of a water supply and wastewater disposal systems for the settlements of Goly Mys, Krolchatnik, Arkhirejka and Centralnaya Usadba, and provision of a water supply for Goly Mys and Krolchatnik by 1 January 2008, and for Arkhirejka and Centralnaya Usadba by 1 December 2009.

Consumer service (work with the consumers) includes:

- developing the principles of direct contractual relationships with the population, to be undertaken jointly with the MHE department
- annual monitoring of people's satisfaction with their water supply and wastewater disposal services

Environmental safety

The environmental safety plan foresees a reduction in untreated wastewater discharges to the Kama river of between 25% and 30% of the total volume of wastewater discharged in 2004, reducing to 3% by 2010 with subsequent eradication of untreated wastewater discharges into the water body.

Leased property

The plan to enable effective use of leased property foresees an improvement (an absence of deterioration) in the indices and parameters for leased property specified in the property acceptance certificates.

d. Evaluation of Novogor-Prikamye, LLC's planned task performance

Performance measurements for Novogor-Prikamye's work involves achieving efficiency measures within key areas specified in Section III of the planned task. The procedure for assessing performance of the planned task will be developed and approved by the coordination committee within six months of the date when the planned task receives assent.

4.5.7 Operational principles of Novogor-Prikamye

The organisation's main principle is of responsibility to consumers – the citizens, enterprises and organisations of Perm - all those that use its water supply and wastewater disposal services:

- its water will be pure, its supply uninterrupted, its wastewater treatment of good quality, its customer service civilised
- the price will be affordable for customers, in relation to which the organization must strive to constantly reduce the cost of its services
- the company must understand that its monopoly position only reinforces its obligations to cus-

tomers

The second principle is responsibility to its employees:
• work will bring satisfaction and protection to employees
• pay will be sufficient and compatible with the contribution made, management will be fair, the length of a working day will be reasonable and work places will be tidy
• each employee will be considered as a personality with the potential capacity for growth
• managers at all levels will be decent, patient and wise, competent and initiative-taking, they will have common sense and understanding, will listen to employees, and will try not to dominate but to lead the way

The third principle is responsibility to business partners:
• the company is obliged to be precise, accurate and willing to oblige in its work with its business partners, striving to set long-term relationships in favour of its customers
• the company's purchases are transparent, they will be based on accreditation, bids and tender procedures
• the company should give preference to local suppliers, if their goods and services are competitive in comparison with those from other regions

The fourth principle is of responsibility to society:
• the company is open and available to the mass media, public organisations and all citizens; it will try to foster publicity about the decision-taking process for all important decisions relating to water supply and wastewater disposal
• the company must protect the environment and do its best to reduce the pollution it creates
• the company's staff must remember that their children and grandchildren will use the fruits of their work

The fifth principle is responsibility to the authorities of the region and territory, both to the executive and representative authorities:
• the company is aware of the degree of its responsibility to the electorate, including the quality of its public utility services
• the company must maintain and develop the municipal property it uses in a proper way, pay a fair tax burden, and take part in city development as a whole
• the company must report on its activities to the authorities

The sixth principle is responsibility to owners and investors:
• business should bring in the necessary income
• it is necessary to develop new capacities, to purchase new equipment, to develop the service infrastructure and to perfect the technologies used
• the company must collect all payments for services rendered
• owners and investors receive a fair remuneration

4.5.8 Novogor-Prikamye's social responsibility
Novogor-Prikamye, LLC's social responsibility includes complying with social interests in accordance with the following parameters:

a. Being a responsible business partner
• the company works in compliance with legal and ethical business practices in forming its relationships with suppliers and contractors under the principles of strict financial discipline
• the search for and selection of suppliers and contractors will undertaken openly, on the basis of declared tenders and bids. All other conditions being equal, priority will be given to local companies. Information about conditions of tenders and their results will be published at the company offices and in the local mass media.

b. Being a responsible employer
• the company acts in strict compliance with the rules of labour law, and will care about the labour conditions and well-being of its employees
• the company has founded the first Council of Veterans (retired employees) and the develop-

ment of an aid programme for veterans is completed.

• a training programme for Novogor-Prikamye personnel is being developed, and there are plans for some 400 employees of the company to receive training under the advanced vocational training programme this year.

• constant communication between the company's employees and its leaders is carried out by means of its corporate newspaper, Novogor. There is a special column called 'My question to the General Director'.

• since January 2005 a new organisational chart has been introduced at the company. This is intended to help to improve production efficiency and to clarify the responsibilities of each leader, in relation to which the current value of each job position was estimated. Pay increase differentials depend on the functions and importance of each job position within the company.

c. Being a responsible citizen

• the company pays its taxes and carries out its activities in accordance with legislative requirements. Novogor-Prikamye has an obligation to provide honest and open information about incidents on their networks, their reasons, and the terms and conditions of liquidation.

• starting in 2005 the company regularly - once quarter - prepares a report about the quality of the water supplied to customers and publishes it in the mass media.

• the company regularly provides information on strategy updates for Perm's water supply and wastewater disposal systems to 2020 together with the city and regional authorities and also the Social Council for the development of Perm's water supply and water disposal systems. At Social Council meetings the most essential issues relating to Novogor-Prikamye's activities and the development of city water supply systems are discussed at least once a quarter.

d. A participant in social relationships

• the traditional form of company participation to social well-being includes support for various charities. Considering that Novogor-Prikamye's main income is via tariff payments from people and businesses the Company aims to provide support for them within the framework of its professional activities. In 2005 the company donated money to veterans of the Vodokanal and to the municipal housing economy as a whole, as well as to educational institutions training profile specialists.

• within the framework of developing social relationships the company is implementing a programme of cooperation with Perm Technical University in order to obtain recommendations for improvement to the company's work and to prepare potential personnel.

• in spring 2005 the company started regular social monitoring to discover what the opinion of the population was regarding the social situation for water supply and wastewater disposal and changes to the action plan for reorganising the water supply.

4.5.9 Social reporting as a reflection of the company's social responsibility

In order to submit the most representative offering in terms of standard social reporting, foreign specialists recommended that the company use the AA1000 standard as its main template with the GRI standard as an additional aid.

Nevertheless, given that Novogor-Prikamye is not listed on the stock market and is a municipal company, the leadership decided to prepare its own form of reporting focused rather on the customers of its water supply and wastewater disposal services rather than on corporate users. The report will be placed at the company's offices, and provided to mass media, authorities and business partners at request.

4.6 Rosvodokanal, OJSC (RVK)

4.6.1 General information

Rosvodokanal (initially known as Rosvodokanalnaladka) was founded in July 1949 by a Resolution of the government of the USSR to disseminate advanced experience in MHE development within such cities as Moscow, Saint Petersburg, Volgograd, Rostov-on-Don, Nizhny Novgorod and Ufa. Rosvodokanalnaladka became the central repository in the USSR of information on the organisation of water supply and wastewater disposal systems in cities of Russia.

In 2003 Rosvodokanal was reorganised in cooperation with Alfa-Eko, and became part of the Alfa-Group.

At present Rosvodokanal:

• has seven engineering centres in Khabarovsk, Ufa, Novosibirsk, Volgograd, Rostov-on-Don, Saint Petersburg and Moscow
• has successful experience in reforming and perfecting MHEs' technical systems, developing and implementing investment projects and raising investment
• undertakes asset management of MHEs in cities with a total population of over two million (Orenburg, Barnaul, Orsk and others)
• provides engineering and consulting services, undertakes engineering surveys and technical audits of MHE systems belonging to the Municipalities of Russia
• develops technical regulations, and justifies operational standards and water supply and wastewater disposal balances for MHE systems
• undertakes design of investments for developing, updating and reconstructing MHE systems
• carries out market assessments of MHE asset values
• makes calculations and provides independent expertise on economically-sound tariffs for water supply and wastewater disposal services

4.6.2 Mission, vision and objective

Vision: to make water available and establish the company as one of the three leaders in Russia's MHE sector.

Mission: the organisation of effective, reliable water supply and wastewater disposal systems on the basis of commercial principles and the safeguarding of public interests. Raising the living standards of Russia's citizens.

The objective pursued by Rosvodokanal is the repair of the sector's damaged relationships and making this process real and perceived by financing scientific and research work, and the project and adjustment work needed to design investments. This will help Rosvodokanal to fulfil the tasks of effective MHE reform in Russia.

4.6.3 Rosvodokanal-Consulting

Rosvodokanal-Consulting is a consulting company established with the participation of foreign partners and accredited by Rosvodokanal, which provides engineering and consulting services. Rosvodokanal-Consulting has a network of regional enterprises with centres in the areas where Rosvodokanal is located, and extends this network.

The locations in which Rosvodokanal and RVK-Consulting work coincide with the location of scientific, educational, project and adjustment centres for the water supply and sewerage facilities (WSSF) of Russia, as they were in the Soviet era.

4.6.4 Projects

At the present time (data as from the end of 2005) Rosvodokanal is implementing the following main projects:
• WSSF reform for Orenburg
• WSSF reform for Barnaul
• WSSF reform for Orsk
• WSSF reform for the Kaluga region
• WSSF reform for Tumen
• WSSF reform for Kazan

4.6.5 MHE reform in Orenburg

Project status: a subsidiary of Rosvodokanal has been managing WSSF Orenburg since 1 December 2004 on the basis of a long-term lease agreement – 20 years, based on the management results during 2003 to 2004. A short-term investment project has been implemented, an assessment of the considerable scope of the engineering works to be undertaken has been carried out and development of a long-term investment programme is in progress.

The results of the work are:
• growth in collection of payments against bills issued from 85% to 98%
• measures to optimise the company's energy use performance taken since March of 2004 have provided a sustainable trend towards reducing energy consumption. The total reduction in energy consumption for 2004 is 5.3%
• the number of failures in the networks have been reduced to a quarter of the original figure, uninterrupted functioning has been achieved in all systems

• water losses have been reduced to 21%
• a reorganisation of the company has been undertaken, to separate the large divisions
• a budget system has been introduced for the whole company and also for each division
• a department of financial planning and control has been set up
• the labour compensation fund and a review of the scope of work have both increased because of reductions in non-manufacturing costs
• an inventory and certification of WSSF property in the city has been undertaken

4.6.6 Rosvodokanal suggestions for reforms to Russia's water supply and sewerage facilities (WSSF)

Rosvodokanal suggests proven decisions regarding reform of unitary enterprises on the basis of Russian and foreign experience should be adopted.

The role of the private sector should depend on the type of partnership, as shown in Figure 4.6.6(a).

Figure 4.6.6(a). Partnership option as indicated by Rosvodokanal experience.

Service Contracts → **Management Contracts** → **Lease** → **BOT, Concession/ Investment Contract** → **BOO/Sale**

The main stages of WSSF reform as shown in figure 4.6.6(b).

Figure 4.6.6(b). The main stages of water supply and sewerage facility reform.

1st stage
Signature of the Agreement and registration of the Company
2-3 months

→

2nd stage
Transfer from MUE to an operator company, Introduction of more efficient management. Short-term management.
Up to 1 year.

→

3rd stage
Signature of long-term lease, investment contract and implementation of investment program
25-49 years

Description of the stages:

First stage:
• creation of a signed agreement about the optimisation of the water supply and wastewater disposal system belonging to Municipality and the registration of a new water supply company

Second stage:
• securing land rights, a technical inventory of the property, its certification and registration with the judicial authorities
• settlement of accounts receivable and payable to the MUE
• a technical audit of all water supply and wastewater disposal systems in order to develop a water supply master plan and strategy for WSSF development
• development of a budget system for the operational activities of the whole organisation and its key departments and preparation of a balanced income and expenditure budget for the organisation and its structural departments;
• implementation of a programme to reduce expenses by optimising the WSSF's management and operation
• measures to increase revenue
• development and approval of a strategic plan for the Vodokanal's corporate development, a long-term investment programme and tariff policy

Third stage:
• conclusion of a long-term lease contract
• implementation of a strategic development plan and investment programme
• improvements in the quality of the services rendered
• raising of investment

4.6.7 Financing sources for the investment programme, in which Rosvodokanal participates
See Figure 4.6.7.

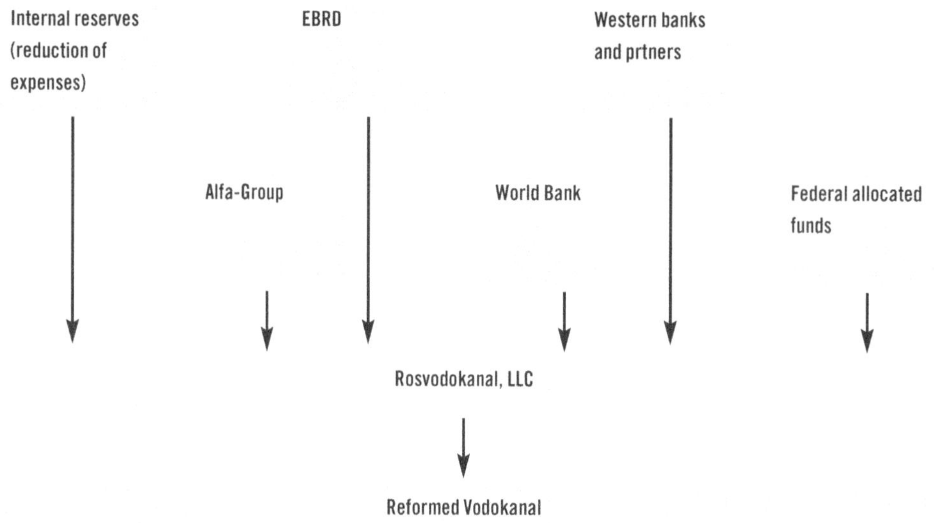

Figure 4.6.7. Financing sources for the investment programme in which Rosvodokanal participates.

5. Perspectives on the reforms to Russia's water supply and sewerage sector

5.1 Assessment of the general situation

At present:

- the federal-level revision of the regulatory and legal framework for the Vodokanals' production activities has been finalised, on the basis of the approaches and standards agreed by the world community
- new approaches to tariff regulation for water supply and sewerage services have been adopted and implemented at federal and regional level, not on the basis of the cost principles of regulation, but on the basis of the interrelationship between cost and the quality of services rendered
- the first projects transferring water supply and sewerage assets into private sector management and the move to increase the private sector's share in such asset management proves the interest of the authorities and the businesses in extending the share taken by the private sector in this process
- in the process of implementing the first projects involving private sector participation in the management of municipal infrastructure assets proof of the availability of significant savings has been demonstrated, in that it has been seen that initial operational cost reductions can be achieved from improvements to the management efficiency and motivation within the Vodokanal's business administrations, rather than from large, long-term investments in the infrastructure
- large-scale corporatisation of the municipal unitary enterprises of the Vodokanals has begun
- the management staff of the Vodokanals that now have private sector participation have received practical experience in effective management within the new economic environment

Nevertheless, the constraining factors for private sector participation in the management of municipal infrastructure assets are:

- at present in Russia there is practically no possibility of obtaining bank credits for long-term investments in municipal infrastructure without related budget guarantees. Investors are not yet ready to invest their funds in the municipal infrastructure in significant amounts (the investment risks are quite high, and such investments do not increase the capitalisation of the investor's company, that is, the municipal infrastructure assets, in which the funds are invested, are not owned by the investors). The assets of those operators that have private sector participation, which manage municipal water supply systems, are not yet sufficient to allow them to obtain credits without additional guarantees.
- there is no practice in real market assessment of real-estate assets within the municipal infrastructure. Engineering infrastructure assets do not have real market values as they are encumbered with the need to be used directly for their purpose and their cost is only estimated, not connected with the real estate market.
- high political risks (the setting of tariffs for water supply and water disposal services is still very much dependent on the individual representatives of the regulatory body, who take the decisions during the tariff approval process).
- the current legislation foresees the possibility at a federal level of setting a maximum limit for tariffs for water supply and wastewater disposal services at the local level, notwithstanding the real investment needs for upgrading and developing the municipal infrastructure (that is, by the investor, even when tariff agreements can be struck with the local regulatory authorities there are no guarantees securing a return on investment and the expected level of profitability).
- when the majority of the factors influencing the attractiveness of investment are uncertain, it is

impossible to make highly-reliable long-range forecasts and financial models

5.2 Short-range forecast

The short-range forecast (up to 2010) is based on the following assumptions:

• the Vodokanals of Moscow and St Petersburg will be corporatised on condition that the controlling block of stock remains with the administrations of these cities

• the Vodokanals in cities with populations of over 500,000 will be reorganised according to plan, whereby an investor receives the controlling block of stock of the operator immediately (at the moment of corporatisation) or shortly after (not later than one year after corporatisation - the French model). Engineering infrastructure will be transferred to the operator's lease on condition that the annual investment volume is assumed to be 10% to 15% of the Vodokanal's annual turnover. This amount of investment volume will be mainly achieved not out of an investor's own funds, but from reducing operational expenditure and improving management efficiency (which increases the attractiveness of investing in such projects).

• Vodokanals in cities with populations from 200,000 to 500,000 (mainly administrative centres of the RF's constituent parts) will be reorganised mainly along the lines of a joint stock company establishment with 100% of the capital in Municipality ownership (the German model). In some cases (really exclusions) these Vodokanals will be transferred on a short-term basis (up to five years) into the management of an investing company with subsequent extension or termination of the management contract.

• Vodokanals in cities with populations from 50,000 to 200,000, in the overwhelming majority of cases, will be corporatised with the Municipality, leading city companies and the Vodokanal management participating in its authorised capital (this is the most profitable variant of Vodokanal reorganisation for the administrations of the Municipalities).

• Vodokanals in cities with populations up to 50,000 will be corporatised mainly by transferring all of their engineering infrastructure into the ownership of a joint stock company (which is included in the authorised capital) and the simultaneous or subsequent transfer (sale) of shares to private investors (the English model). In some cases (which will be quite rare), the shares of such joint stock companies will remain the property of the Municipality. Taking into consideration the low percentage of household expenditure spent on water supply and water disposal services it would be more profitable for the administration of a small city to find an investor who would take the municipal structure assets into ownership, raise investments to improve them and provide effective management for these systems, compared to financing such improvements through a budget (which is unlikely in practice to produce external long-term investments in small cities, if the municipal engineering infrastructure remains in municipal ownership).

As a result of the reforms, by 2010 the following actions will have been undertaken in the Russian market for water supply and wastewater disposal:

• the creation of a modern regulatory framework will have been completed; this framework will take into consideration international experience and will regulate all areas of Vodokanal activity

• five to six main national management companies will be actively functioning and developing; these operators will share at least 50% of the market for water supply and water disposal services in line with territorial principles (EVP – Siberia, RVK – the west and south of Russia, Novogor-Prikamye – the North of Russia and the Urals, Mosvodokanal – the centre of Russia, the Vodokanal of St Petersburg – the north-west of Russia). Water supply and sewerage systems for cities with populations of over 500,000 will be transferred to these companies into leases or management, and at the same time the ownership of engineering infrastructure assets will remain with the Municipalities; (probably, in some cases, these companies will take all of the water supply and sewerage systems in the region as a 'uniform load')

• the majority of the administrations of the Municipalities in cities with populations up to 500,000 will learn to evaluate the actual value of the Vodokanals' assets that they own and will be able to find the most profitable strategy for managing these assets (to sell them at their actual market value or to manage them independently, receiving income from income gain dividends)

• because of increased information and communications possibilities, by means of the internet interested consumers will be able to find and get all the information they need in order to compare management results from the municipal water supply and wastewater disposal systems in their city to other cities on the internet. As a result of such comparisons they will be able to influence Vodokanal activities in their city through their chosen local authorities.

6. Conclusions

6.1 Concluding remarks

The authors of the present report have attempted an analysis of the condition of the municipal water supply system in Russia at the beginning of the 21st century and to generalise about the relatively small (only two to three years') experience of reform in Russia's water supply and wastewater disposal sector.

The involvement of the private sector in Russia's municipal water supply and sewerage services market and the integration of the country into the global economy dictate the necessity of using the same laws as are used by the world community. The world community considers it absolutely natural that a Vodokanal cannot undertake glassmaking, and that provision of water supply and wastewater disposal services is a reliable and profitable business. Though the world's drinking water resources are quite large in most cases they are difficult to access, which is why there will be growing attention on this sector of the economy.

Because of these factors, it is very likely that the participation of the private sector in Russia's public water supply and wastewater disposal services will quite soon become as natural as its participation in the production of bottled water.

The most important issue for a consumer is not which company provides its services (whether private or state). The most important thing for a consumer is how he or she can influence this company in the event that its services are not satisfactory or if it overcharges. Global experience proves that this can be achieved by effective state regulation of private water supply companies, ensuring transparency in their operations and transactions, and motivating the consumers themselves and fostering their participation in discussing and developing strategic solutions for the public water supply and wastewater disposal services in their own city, region, and country.

Appendix 1

Constituents of the Russian Federation

Central Federal District (18 constituents)
Regions (17): Belgorod, Bryansk, Vladimir, Voronezh, Ivanov, Kaluga, Kostroma, Kursk, Lipetsk, Moscow, Oryol, Ryazan, Smolensk, Tambov, Tver, Tula and Yaroslavl.
Federal City (1): Moscow

North-West Federal District (11 constituents)
Republics (2): Karelia and Komi.
Regions (7): Arkhangelsk, Vologda, Kaliningrad, Leningrad, Murmansk, Novgorod and Pskov.
Federal City (1): Saint Petersburg
Autonomous national area (1): Nenetsk

Southern Federal District (13 constituents)
Republics (8): Adygeya, Dagestan, Ingushetia, Kabardino-Balkarian, Kalmykia, Karachyevo-Cherkess, North Ossetia and Chechen.
Territories (2): Krasnodar and Stavropol.
Regions (3): Astrakhan, Volgograd and Rostov.

Volga Federal District (14 constituents)
Republics (6): Bashkortostan, Mariy-El, Mordovia, Tatarstan, Udmurt and Chuvash.
Territory (1): Perm.
Regions (7): Kirov, Nizhny Novgorod, Orenburg, Penza, Samara, Saratov and Ulyanovsk.

Uralian Federal District (6 constituents)
Regions (4): Kurgan, Sverdlov, Tumen and Chelyabinsk.
Autonomous national areas (2): Khanty-Mansi and Yamalo-Nenetsk

Siberian Federal District (16 constituents)
Republics (4): Altai, Buryatia, Tyva and Khakassia.
Territories (2): Altai and Krasnoyarsk.
Regions (6): Irkutsk, Kemerovo, Novosibirsk, Omsk, Tomsk and Chita.
Autonomous national areas (4): Aginsky Buryatsky, Taimyr, Ust-Ordynsky Buryatsky and Evenki.

Far East Federal District (10 constituents)
Republics (1): Sakha (Yakutia).
Territories (2): Primorsk and Khabarovsk.
Regions (4): Amursk, Kamchatka, Magadan and Sakhalin.
Autonomous region (1): Jewish.
Autonomous national areas (2): Koryak and Chukotka.